William Lloyd Baily, Spencer Fullerton Baird

Our own Birds

A Familiar Natural History of the Birds of the United States

William Lloyd Baily, Spencer Fullerton Baird

Our own Birds
A Familiar Natural History of the Birds of the United States

ISBN/EAN: 9783337330910

Printed in Europe, USA, Canada, Australia, Japan

Cover: Foto ©berggeist007 / pixelio.de

More available books at **www.hansebooks.com**

Bald Eagle. (*Frontispiece.*)

PREFACE.

THE object of this book is not to treat the subject of Ornithology scientifically, but simply to present in a concise and familiar manner to the youthful reader, some interesting facts relating to the birds of our own country. Various works have been written and published upon this subject, containing, probably, all that the student or amateur could wish to know; but being both voluminous and expensive, they are quite beyond the reach of children. They also contain, in connection with a variety of interesting matter, an array of scientific details, which to most young persons are unintelligible, and which can only be appreciated by the more advanced student. We have, therefore, while adhering strictly to an approved systematic arrangement of the Genera and Species, endeavored to avoid, as much as possible, the use of all terms and expressions which would in any degree confuse the reader, or detract from the

1*

interest of the work, — hoping thereoy to excite in some a degree of love for a study, which they will find to be at once entertaining and instructive, as well as conducive to the health of body and mind.

We have confined our descriptions chiefly to the Birds of the United States, but in a few instances have introduced others for the purpose of better illustrating the subject, or increasing our information respecting the peculiarities of any tribe.

It is proper to observe that, while many of the cuts in the following pages are original, others are after Audubon.

CONTENTS.

CHAPTER I.

INTRODUCTION.

CHAPTER II.

INSESSORES: *PASSERES — OSCINES.*

(vii)

CHAPTER III.

INSESSORES: *PASSERES — OSCINES.*

CHAPTER IV.

INSESSORES: *PASSERES — OSCINES.*

CHAPTER V.

INSESSORES: *PASSERES, CLAMATORES, AND OSCINES.*

CHAPTER VI.

INSESSORES: *SYNDACTYLI.*

CHAPTER VII.

INSESSORES: *SYNDACTYLI AND ZYGODACTYLI.*

CHAPTER VIII.

INSESSORES: *SYNDACTYLI.*

CHAPTER IX.

INSESSORES: *ACCIPITRES.*

CHAPTER X.

INSESSORES: *PULLASTRÆ.* CURSORES: *GALLINÆ.*

CHAPTER XI.

CURSORES: *GRALLÆ*.

CHAPTER XII.

NATATORES.

OUR OWN BIRDS.

CHAPTER I

INTRODUCTION.

DESCRIPTION OF THE DIFFERENT PARTS OF BIRDS—CLASSI-
FICATION: RAPTORES, INSESSORES, SCANSORES, RASORES,
GRALLATORES, NATATORES—ON THE FLIGHT OF BIRDS—
THEIR PLUMAGE, INSTINCT, MIGRATION, NESTS, EGGS, GE-
NERIC DIVISIONS.

IT may be said that there is no part of the Animal
Kingdom in which a more general interest is felt
than in Birds. The great variety of their forms, the
beauty, and often the gaudiness of their plumage,
their graceful motions, their peculiar habits and manners, and, above all, their sweet musical voices, all
conspire to assign them a most prominent position in
Nature's parterre.

The birds of our own country, although less brilliantly attired than some others, must yet hold in our
affections the foremost place. What happy associa-
tions do we connect with them! Who that listens to

(11)

the Cuckoo's voice, thinks not of his boyhood, when, thoughtless of time's passing wing, he has stopped by the wayside, and watched her building her nest? Who that hears the song of the Blue-bird and Linnet, finds not in their sweet notes a tie that binds to his heart some memory of the past? and is ready to exclaim:

> "And I can listen to thee yet
> And lie upon the plain;
> And listen till I do beget
> That golden time again."

Birds are ever around us: — their busy active life displays itself wherever we turn our steps :—even at those seasons when most species have retired to the sunny south, a few still remain to cheer our hearts and enliven our homes. But it is in the spring and summer that we become most familiar with these feathered tenants of the air. When the clouds of winter, and its lowering storms, have rolled themselves behind the hills, — when the sun shines out with renewed warmth and vigor, and the softened breath of Heaven wafts from the flowery fields and leafy woods a pleasing fragrance, the Blue-bird, the Song Sparrow, and the Robin, with thousands of lovely comrades, fresh from their winter haunts, come again to cheer us with a welcome music. The Swallows twitter gaily as they sail over the meadows; the Wren, perched upon a neighboring twig, sings to his mate while she turns from her accustomed box the remains of last year's nest; the busy little Warblers and Fly-catchers, incessantly active, are plying their bills voraciously among the insect life; the Hawk wheels his buoyant flight in graceful circles overhead; and the Humming Bird darts like a meteor in pursuit of some favorite flower. All these cast a halo of attraction around the name of Spring, reminding us that "the time of the singing of birds is come."

Before entering fully upon our subject, there are a few observations which it will be necessary for us to make, which cannot but be of use to the young student of Ornithology. How wonderfully is the form of a bird adapted to the element in which it is designed to move! We perceive that the general outline of its body is boat shaped, as being most consistent with a rapid motion through the air. The skeleton is admirably formed, both as to strength and buoyancy, many of the larger bones being hollow, and filled with air instead of marrow. In the development of the muscles, also, we see compactness combined with prodigious force; and the structure of the wing at once commends itself to our notice as a propelling, as well as a supporting power, peculiarly suited to the wants of the bird; while the whole body is clothed with a covering which, for lightness, warmth, and beauty, could hardly be excelled.

By reference to the adjoining cut, the following description of the different parts of a bird, and the names applied to some of the most prominent feathers, will be understood:

2

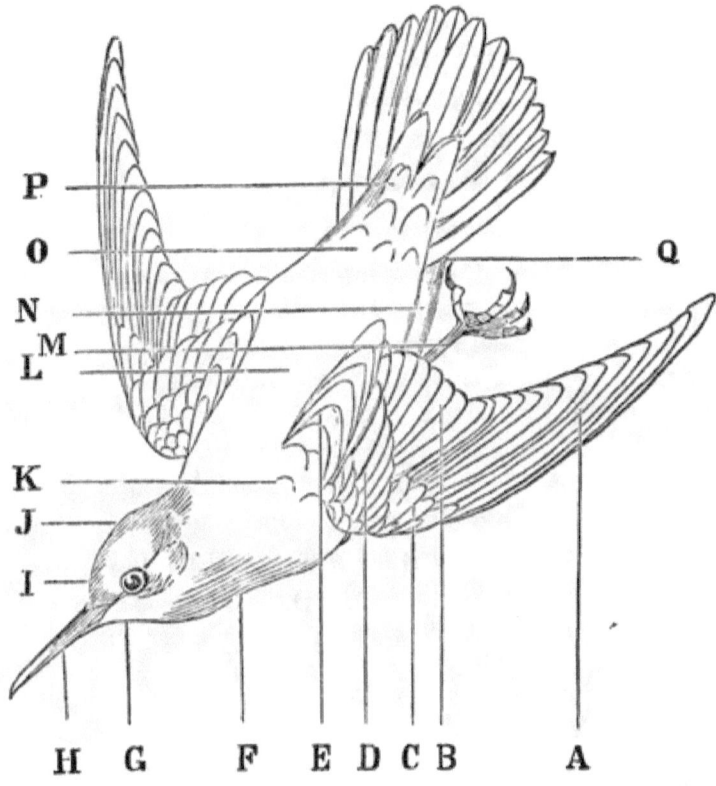

A. Primary quills.
B. Secondary "
C. Spurious wing.
D. Greater wing coverts.
E. Tertiary quills.
F. Throat.
G. Jugulum.
H. Bill.
I. Front.
J. Crown.
K. Scapular feathers.
L. Interscapular.
M. Tarsus, or leg.
N. Abdomen.
O. Rump.
P. Upper tail coverts.
Q. Lower " "

Birds have been by some naturalists divided into six different orders, as follows :

1st.	ACCIPITRES.	(*Preying*).	Eagles, Falcons, and Vultures.
2d.	INSESSORES.	(*Perching*).	Sparrows, Warblers, Thrushes, &c.
3d.	SCANSORES.	(*Climbing*).	Woodpeckers and Parrots.
4th.	GALLINÆ.	(*Scraping*).	Pheasant, Partridge, &c.
5th.	GRALLÆ.	(*Wading*).	Heron, Crane, and Ibis.
6th.	NATATORES.	(*Swimming*).	Geese and Ducks

Each order, it will be seen, possesses a peculiar formation of the bill, wings, or feet; and it is by the close observance of them, as well as of differences in their plumage, that the naturalist is enabled to distinguish between the different species.*

---

* With fuller anatomical information upon the subject, later zoologists have regarded the following arrangement as most nearly representing Nature:

Of the class Aves there are three sub-classes, viz., Natatores, (principally aquatic), Cursores, (principally terrestrial), and Insessores, (principally arboreal).

The sub-class, NATATORES, embraces four orders, viz., the *Pygopodes*, (containing four families; grebe, loon, penguin, etc.); the *Longipennes*, (two families; petrel, gull, etc.); the *Steganopodes*, (two families; pelican, etc.); and the *Lamellirostres*, (two families; ducks, mergansers).

The sub-class, CURSORES, consists of three orders; first, *Grallæ*, (containing six families, rails, herons, flamingo, snipe, plover, etc.); second, *Brevipennes*, (two families, ostrich, apteryx, etc.); third, *Gallinæ*, (four families, grouse, pheasant, turkey, etc.)

The sub-class, INSESSORES, is a union of five orders. First, *Pullastræ*, (four families, brush turkey, dodo, pigeon, etc.); second, *Accipitres*, (three families, the birds of prey); third, *Syndachyli*, (seven families, hornbill, kingfisher, humming bird, swift, whip-poor-will, etc.); fourth, *Zygodactyli*, (seven families, trogons, cuckoos, woodpeckers, parrots, etc.); fifth, *Passeres*, (twenty families, sparrows, thrushes, tanagers, crows, etc.) The third of these orders is of uncertain limits; very good authorities refer it to the fifth, (Passeres), forming from part of it a sub-order, (Strisores). In the system here sketched, the Passeres comprise two sub-orders, distinguished partly by the greater or less perfection of the vocal organs. They are the *Clamatores*

In the Accipitres, represented by the Eagles and Falcons, the wings are long and powerful, and their food consisting mostly of the flesh of small animals, they are not only assisted in their pursuit of them by a rapid and vigorous flight, but the form of the feet and claws is such as to enable them to seize and secure their prey, while the hooked beak is well suited to the purpose of tearing it in pieces.

The Insessores embraces a great variety of birds exhibiting a corresponding variety of form. A large majority of them feed upon insects and their larvæ or eggs; and while in all, the feet are well adapted for perching, the bill and wings will be found to vary much according to the habits of the bird. The Swallows, Fly-catchers, Tyrants, etc., pursue their food upon the wing; they have therefore great powers of flight, the mouth is wide, the bill broad at the base, and sometimes armed at the extremity with a slight hook. The Warblers, Thrushes, Wrens, and many others, seek their food among the branches and leaves of the trees, feeding mostly upon the worms,

---

and *Oscines*. Of the ten families belonging to the first (the inferior), five are represented in the United States; of the second, which exhibits the higher organization, the whole ten families exist in our country.

This system was first published at Upsal in 1860, by Wilhelm Lilljeborg.

In the present book, the liberty has been taken of altering the author's arrangement as far as possible, and the classification proposed by Lilljeborg has been substituted.

E. D. C.

the chrysalis, or the eggs. These are possessed of extraordinary agility in hopping about from twig to twig in search of food. Some birds of the order Incessores live on seeds and nuts; such are furnished with a strong short beak, quite thick at the base; the two mandibles sometimes working together like a pair of scissors. To this class belong the Finches, Sparrows, Crossbills, with many more.

The third division, Scansores, or Zygodactyli, comprises the family of Woodpeckers, Cuckoos, Parrots, etc. In this division the arrangement of the toes is peculiar, two before and two behind, which enables the bird to grasp with a firmer hold the bark or the branches, while climbing from one part of the tree to another. To the Woodpecker this arrangement is of peculiar service, almost its whole life being spent in clambering over the rough surface of trunks and branches of trees.

The case is much the same with the Parrots, although their climbing propensities are confined more to the smaller branches, in which they make good use of their strong hooked beak, hanging on with it, while taking a fresh foothold.

In the fourth division, Rasores, we will notice a marked change, the whole bird differing widely in form and appearance from the preceding; the body becomes larger and less buoyant, the wings less ample, and consequently the flight restricted, the bill adapted to picking up seeds and berries or to the cropping of tender herbage, while the feet are formed for walking on the ground and the claws for scratch-

2 *   B

ing in the earth. By observing the contour of the Turkey and the Pheasant, or of the common poultry of our barn-yards, it will be seen that they are formed for a terrestrial existence, and that their heavy bodies and less capacious wings unfit them for much aërial locomotion.

In the Grallæ, or fifth division, a long bill, long neck, long legs, and sometimes long toes, are the prominent features. The Heron, the Crane, the Curlew, and others of this class, often seek their food in deep waters, into which they wade as far as the length of their legs will permit, and, with the head resting upon the shoulders, they stand silently, with their eyes fixed upon the stream, until some unwary fish comes within their reach, when they dart out their long necks and catch it.

The Natatores are a large order, composed of Ducks, Swans, and Geese; these live almost exclusively in the water; they are web-footed, and swim very beautifully, sailing about on the surface like a miniature ship. The bill is of a peculiar formation, being broad and somewhat boat-shaped, and rounded at the extremity. They feed upon the vegetation found growing in, and on the margin of, the water, also upon worms, larvæ, etc. Although these birds seem more adapted for a life divided between the land and the water, yet they are possessed of great powers of flight, and are often seen in considerable numbers soaring aloft and progressing with a very rapid motion. Their line of march is singular, one generally taking the lead, and the rest following in single file, either in a straight line or in the form of the letter V.

To the student of Ornithology, the flight of birds, and the motion of the wings peculiar to the different tribes, will form an interesting subject for observation. To the practised eye, this is quite a sure indication of the class to which the bird belongs.

By those who are familiar with the easy and unrestrained flight of the Eagle, he is at once recognized. Now he soars in graceful curves at an immense height, as though intent on viewing the whole earth beneath him,—then with unmoving wing glides in a horizontal course until lost in the deep blue vault of Heaven. The motions of the Turkey Vulture are also of a most singular and interesting character. These birds may often be seen sailing overhead for hours together, moving in curves or gently undulating lines, rising and falling at pleasure, with but little apparent motion of the wings, and sometimes ascending in easy circles beyond the reach of vision.

The Woodpecker describes, in its course through the air, a waving line, which is in consequence of the wings being alternately closed and expanded at intervals during flight. The Sparrows also perform a zigzag course, rising and falling first to the one side and then to the other. In the Fly-catchers the motion of the wings is rapid and steady; sometimes in long-continued flight their course is slightly undulating. The Humming Bird darts with the swiftness of an arrow, and the vibrations of its wings are so incessant as to render them almost invisible; while the Heron and the Crane wheel their heavy bodies through the air with a slow but steady flapping of a pair of ample, curving wings, their heads drawn in toward the body, and their long legs following like a rudder.

It is very evident that the shape of the wings, and the arrangement and texture of the feathers composing them, must have a material effect upon the flight of birds. A long, pointed, flat wing, with stiff and close-set primaries, is undoubtedly best adapted to rapidity of motion. This will be most observable in the Swallow, the Humming Bird, and the Night Hawk, which of all birds are the most remarkable for the nimbleness and agility of their movements. How beautifully does the Swallow skim over the meadows and lakes, or mount aloft in the air, now wheeling to the one side and then darting like an arrow to the other! And how graceful are the antics of the Night Hawk as he pitches his aërial summer-sets, or gambols with matchless ease across the sky!

It will be observed that the wings of birds of rapid flight are seldom very concave beneath; on the contrary, they are generally quite flat when extended. This flatness, although it contributes to the velocity of motion as the bird sweeps along, destroys to a great extent the power of direct ascent. Where the wing is of moderate length and concave, as in the Owl, and composed of loose soft feathers, the flight is buoyant and noiseless, and quite different from that of the Falcon, the feathers being too soft and yielding to produce any whistling or rushing noise. A short, rounded, concave wing, is mostly peculiar to birds of terrestrial habits, as it will at once be seen that this form is least adapted to extensive progress through the air. The wings of the Partridge and Pheasant are of this shape.

Appendages of various kinds are occasionally attached to the wings of birds: — the direct uses of these cannot readily be ascertained. We must therefore conclude that they were designed rather as ornaments than to minister to the comfort or convenience of the bird. In the Leona Night Jar, a bird allied to the Night Hawk, and a native of Africa, from the centre of the upper wing coverts issues a slender flowing shaft about twenty inches in length, and tipped for about five inches with a broad web. In some the scapularies are elongated into delicate and graceful plumes, as in the Heron and Crane.

While, as has been shown, most birds possess the power of flight in a greater or less degree, yet there are a few species to which it has been wholly denied. This is in consequence of two separate peculiarities in the development of those organs which are so nicely adapted to their aërial habits. In the Ostrich and Emu we see merely the rudiment of a wing, destitute of the ordinary bony and muscular structure; and in the Penguin and Auk, the wing, although possessed of considerable muscular power, is converted into an organ of aquatic progression, and is covered with close, stiff, and scale-like feathers.

The tail also exerts considerable influence in guiding the motions of the bird through the air, acting as a rudder to direct its course, and it also assists greatly in preserving a proper equilibrium, both in motion and while at rest. The form of the tail differs widely in different species; perhaps there is no other part of its plumage in which so great a diversity exists, and often the male and female are so unlike in this respect as scarcely to be recognized as being different sexes of the same bird.

The structure of a simple feather is in itself a wonder, — its unique form, its soft and delicate texture, its perfect adaptation to the use for which it was designed,—the amazing difference which exists between those of different birds, from the stout quill of the Buzzard's wing to the shining spangle from the Humming Bird's throat, the plain but exquisite shadings and markings of the one contrasted with the gaudy and glittering hues of the other, display the infinite wisdom and the matchless skill of Him who is

"Wondrous alike in all he tries!"

The male bird is *mostly* clothed in more brilliant plumage than his mate, and the young of both sexes generally assume the garb of the female until the following spring. Thus it appears that color not only serves the purpose of beauty, but also of protection, for while the gay adornment of the male attracts the attention and makes him a more certain mark for the sportsman, the female to whom is committed the care of the young is secured from danger by her unobtrusive dress.

The Partridge and Woodcock, which mostly live

upon the ground, are secreted from the searching eye of the Hawk and the Kite by their grey speckled plumage, which resembles the ground on which they move. The tawny feathers of the Whip-poor-will also afford it a means of protection, even from man, as it is extremely difficult to distinguish it from the log upon which it may be crouching, almost within our reach. The Ptarmigan, which inhabits very cold northern climates, in summer has its plumage marked with stripes of black or brown, which colors more nearly approach to those of the rocks and barren heaths upon which it lives; but, did these hues remain during the winter, when the snow covers every object with a mantle of white, the place of its concealment would be readily discovered, and it would fall an easy prey to the Snowy Owl or the Gyr Falcon. What then is the provision of Nature to guard against this danger? As the cold season advances, the feathers, by some unknown process, gradually become white, and the bird burrows fearlessly in the snow, in search of berries and leaves, comparatively secure from the eye of its enemy.

Another object besides safety is gained from the concealment afforded by the peculiar colors and markings of the plumage; the support of the bird being sometimes dependent upon it. Thus the Crane and the Heron, and many other water birds, which depend upon their dexterity as fishers for their supply of food, are clothed with feathers partly of white and partly of a bluish slate color, and the fish as they glide beneath the water recognize but little difference

between the plumage of their foe and the blue heaven above them studded with clouds, and passing on
fearlessly, they fall an easy prey to his voracious appetite, while did the bird present a darker image
against the sky, it would produce alarm, and the fish
would hurry off to the protection of some overhanging bank, or dive into the depths below.

The instincts of birds are in many respects very
remarkable. What sagacity do they display in dis-
cerning the proper time for performing their migrations! With what precision do they direct their
course through the darkness of night! With what
skill do they construct their nests! And with what
tender affection do they provide for the wants and
protection of their young! It has been observed of
the House Wren and many other birds, that the same
pair will return to the same spot for many successive
seasons: that these little creatures should be able to
designate in a journey of at least one thousand miles
the precise spot where they have nestled the year before, shows a degree of intelligence not always found
even in man.

The migration of birds is by no means the *least*
interesting part of their history. How often do we
observe when looking out upon some bright morning
of spring, that while the air seems laden with fresh
odors, it also bears upon its bosom a soft aërial music,
a sweet incessant warble, the song of thousands of
merry little travellers fresh from the distant south!
Each day for weeks in succession seems to bring new
arrivals, until at last we welcome the tardy little

Humming Bird, which, although swift of wing, is often among the last in the train.

Birds frequently perform their migrations at night, halting at convenient distances, sometimes spending many days in a congenial spot, and only leaving it as the advance of the season warns them of the necessity of completing their journey. The rapidity with which some species travel through the air on these occasions, has been the subject of much speculation. It is a well-ascertained fact that their swiftness is so great as far to surpass any speed which it is in the power of man to produce, and has been known sometimes to be equivalent to one hundred miles an hour. The males generally arrive a few days in advance of the females, as though for the purpose of reconnoitering and finding out suitable places to locate their nests; and the coming of the females is a signal for the choosing of mates, and making general preparations for the accommodation of a family.

That every distinct species constructs a nest of some peculiar shape, and of materials best adapted to its own wants, is a circumstance worthy of notice. The unique little structure built by the Humming Bird from the finest and most delicate moss, and lined with the soft down from different plants, while it is well calculated for the accommodation of its own tiny progeny, would hardly answer the wants of any of its neighbors. The Eagle rears her young upon some bare and inaccessible crag, where, with a heap of sticks and moss for a nest, she broods over them in solitude.

3

The Oriole weaves a neat little bag of bark, fine grass and wool, often strengthened with pieces of string or horse-hair, and hangs it from the twigs of some waving bough, which rocks to and fro in the wind, and there in the midst of a storm which would demolish a structure of greater weight and firmness, she sits at her ease, under the protection of Him, "without whose notice not a sparrow falleth to the ground." The White-eyed Vireo, whose nest is in the shape of an inverted cone, suspends it from the circling stems of a running vine. The Whip-poor-will and the Chuck-wills-widow merely scrape away the leaves near some prostrate log, or among the thick undergrowth of the forest, and lay their eggs upon the bare ground; but so nearly does their color resemble that of the leaves and earth, that it is almost impossible to discover them unless their concealment is betrayed by the flight of the bird.

The number of eggs deposited in a nest varies greatly. The Whip-poor-will lays two, the Partridge from fifteen to twenty-four. The color also varies much in the different species; some are of a deep and beautiful blue, others as white as snow; some are marked with irregular blotches near the great end, or spotted thickly all over with brown on a yellowish or light olive-colored ground; but perhaps the most common color is one uniform speckled mixture of various shades of gray. Mostly but one brood is raised in a season, but frequently two, and with those birds which arrive early sometimes three.

One of the most remarkable instincts of birds is displayed in their keen sense of approaching danger, and in the means which they adopt to avoid it. So sagacious are the birds of some species, that they will always keep at a safe distance from the gun of the sportsman, although they may never have had any experience of the danger of coming within range of his shot. This is particularly the case with the Crow; so cautious are they that when a flock is committing depredations upon the farmer's corn-stacks, they keep a sentinal posted in some elevated position to give notice of the approach of any suspicious looking individual. Young Ducks, almost as soon as they have left the shell, will seek the water, often to the amazement of the hen who has adopted them; here they will swim about and catch gnats and flies; but a wasp they will avoid, as its sting would be injurious. Chickens will show no signs of fear at the approach of a strange turkey or goose; but if a hawk hovers in the distance, they will become agitated and seek shelter. Some birds, if the vicinity of their nests is approached, will immediately fly to the ground before the intruder, and dropping their wings as though wounded, will limp about in great apparent distress; by this means they often deceive those who are ignorant of their habits, and gradually lead them away from the spot, when their assumed lameness suddenly disappears, and they fly off as nimbly as ever.

In order to facilitate the study of Ornithology, the

six great orders of birds have been divided by naturalists into groups or families called *genera*, to each
of which a name has been applied often indicative
of some peculiarity in the appearance or habits of
the bird, and mostly expressed in Latin.   Thus, to
the Sparrow family has been applied the generic term
of *Fringillidæ*, and to the Humming Birds that of
Trochilidæ.   These families are again separated into
sub-genera, according to certain differences in the formation of the bill, feet, wings, etc.   To each of these
divisions a name is given depending much upon the
fancy of the naturalist, and is frequently bestowed
in honor of some great patron of science.   The sub-
genera often consist of many *species*, and an appropriate Latin *specific* name is added to each, by which
it may be distinguished from all others.   Thus, the
common House Wren of Europe is called *Troglodytes
Domestica*—the former being its family or generic,
and the latter its individual or specific title.   Thus
to every little warbler that sings its matin song beneath our windows, science has given a name as significant as John Smith or John Jones, the only difference being that the part of it which designates his
family is mentioned first.

By a careful attention to the foregoing remarks,
and by frequent observation of the habits of birds in
their accustomed haunts, the young student may soon
become acquainted with the appearance and manners
of most of our native species, and with the aid of a
little study will be able to recognize in each a fami-

liar friend, who is ever ready to minister to his pleasure, either by cheering his solitary hours with a lively song, or by abstracting his thoughts from the artificial world about him, and turning them to the contemplation of the wonderful works of an All-wise Providence.

3 *

CHAPTER II.

INSESSORES: *PASSERES — OSCINES.*

DESCRIPTION OF THE BOBOLINK BY "WASHINGTON IRVING"—
COW-BIRD—RED-WINGED BLACKBIRD—BALTIMORE ORIOLE
—BULLOCK'S ORIOLE—ORCHARD ORIOLE—MEADOW LARK
—RAVEN.

THERE are comparatively few persons who are aware of the pleasure that is to be derived from a morning ramble in the woods, for the purpose of observing the habits of the countless feathered beings which abound on every hand. There is a real enjoyment in watching their incessant activity, the beauty and singular ease of their motions; to trace the gaudy colors in which some are clothed, and the plainer though no less pleasing tints of others; to examine the beautiful and delicate structure of their nest; and, above all, to listen to the sweet and mellow cadences of their many-toned voices. A few hours thus spent can hardly have any other than a happy effect. There is one thing which cannot fail to strike us as a prominent feature among Birds, that is, variety — not only in their plumage and song, but in their habits : a variety which ends only with the species, each seeming to possess a distinct character of its own.

Passing from the Finches we come to the Blackbirds and American Orioles, foremost among which is the Bobolink, or Reed-bird or Rice-bird, that bright, active little bird which comes to us in the spring in a beautiful coat of black and white, sings sweetly for

a few short weeks, then changing his suit for one of dusky grey, commences a process of gormandizing which soon fits him for the gun of the sportsman and the epicure's table.

Boboliuk, or Reed-bird.

The following beautiful description of this bird is from the pen of Washington Irving:

"The happiest bird of our spring, and one that rivals the European lark in my estimation, is the Boblincon or Boblink, as he is called. He arrives at that choice period of our year which, in this latitude, answers to the description of the month of May, so often given by the poets. With us it begins about the middle of May, and lasts until nearly the middle of June. Earlier than this, winter is apt to return on its traces, and to blight the opening beauties of the year; later than this begin the parching and panting and dissolving heats of summer. But in this genial

interval Nature is in all her freshness and fragrance;
'the rains are over and gone, the flowers appear upon
the earth, the time of the singing of birds is come,
and the voice of the turtle is heard in the land.'
The trees are now in their fullest foliage and bright-
est verdure; the woods are gay with the clustered
flowers of the laurel; the air is perfumed by the
sweet-brier and the wild rose; the meadows are ena-
melled with clover blossoms; while the young apple,
the peach, and the plum begin to swell, and the
cherry to glow among the green leaves.

"This is the chosen season of revelry of the Boblink. He comes amid the pomp and fragrance of
the season; his life seems all sensibility and enjoyment, all song and sunshine. He is to be found in
the soft bosoms of the freshest and sweetest meadows, and is most in song when the clover is in blossom. He perches on the topmost twig of a tree, or
on some long, flaunting weed, and as he rises and sinks
with the breeze, pours forth a succession of rich tink-
ling notes, crowding one upon another like the out-
pouring melody of the Skylark, and possessing the
same rapturous character. Sometimes he pitches
from the summit of a tree, begins his song as soon
as he gets upon the wing, and flutters tremulously
down to the earth, as if overcome with ecstasy at his
own music. Sometimes he is in pursuit of his par-
amour, always in full song, as if he would win her
by his melody, and always with the same appearance
of intoxication and delight.

"Of all the birds of our groves and meadows, the

Boblink was the envy of my boyhood. He crossed
my path in the sweetest weather and the sweetest
season of the year, when all Nature called to the
fields, and the rural feeling throbbed in every bosom,
but when I, luckless urchin! was doomed to be mewed
up during the livelong day in that purgatory of boyhood, a school-room. It seemed as if the little varlet
mocked at me as he flew by in full song, and sought
to taunt me with his happier lot. Oh, how I envied
him! No lessons, no task, no hateful school; nothing
but holiday, frolic, green fields, and fine weather. Had
I been then more versed in poetry, I might have ad-
dressed him in the words of Logan to the Cuckoo:

> 'Sweet bird! thy bower is ever green,
>   Thy sky is ever clear;
> Thou hast no sorrow in thy note,
>   No winter in thy year.
>
> 'Oh! could I fly, I'd fly with thee,
>   We'd make, on joyful wing,
> Our annual visit round the globe,
>   Companions of the spring.'

" Further observation and experience have given me
a different idea of this little feathered voluptuary,
which I will venture to impart for the benefit of my
school-boy readers, who may regard him with the
same unqualified envy and admiration which I once
indulged. I have shown him only as I saw him at
first, in what I may call the poetic part of his career,
when he in a manner devoted himself to elegant pursuits and enjoyments, and was a bird of music, and

song, and taste, and sensibility, and refinement. While
this lasted he was sacred from injury; the very school-
boy would not fling a stone at him, and the merest
rustic would pause to listen to his strain.   But mark
the difference.   As the year advances, as the clover
blossoms disappear, and the spring fades into summer, he gradually gives up his elegant tastes and
habits, doffs his poetical suit of black, assumes a rus-
set, dusky garb, and sinks to the gross enjoyment of
common vulgar birds.   His notes no longer vibrate
on the ear; he is stuffing himself with the seeds of
the tall weeds, on which he lately swung and chaunted
so melodiously.   He has become a 'bon vivant,' a
'gourmand;' with him now there is nothing like the
'joys of the table.'   In a little while he grows tired
of plain, homely fare, and is off on a gastronomical
tour in quest of foreign luxuries.   We next hear of
him, with myriads of his kind, banqueting among
the reeds of the Delaware, and grown corpulent with
good feeding.   He has changed his name in travel-
ing; Boblincon no more, he is the *Reed-bird* now,
the much-sought-for titbit of Pennsylvania epicures,
the rival in unlucky fame of the ortolan!   Wherever
he goes, pop! pop! pop! every rusty firelock in the
country is blazing away.   He sees his companions
falling by thousands around him.

   "Does he take warning and reform?   Alas, not
he!   Incorrigible epicure!   again he wings his flight.
The rice-swamps of the South invite him.   He gorges
himself among them almost to bursting; he can scarcely
fly for corpulency.   He has once more changed his

name, and is now the famous *Rice-bird* of the Carolinas.

"Last stage of his career, behold him spitted
with dozens of his corpulent companions, and served
up a vaunted dish on the table of some Southern
gastronome.

"Such is the story of the Boblink : once spiritual,
musical, admired, the joy of the meadows, and the
favorite bird of spring; finally a gross little sensual-
ist, who expiates his sensuality in the larder.   His
story contains a moral worthy the attention of all
little birds and little boys, warning them to keep to
those refined and intellectual pursuits which raised
him to so high a pitch of popularity during the early
part of his career; but to eschew all tendency to that
gross and dissipated indulgence which brought this
mistaken little bird to an untimely end."

The Bobolink and the Cow-bird form a small
group which connects the Finches with the true
Blackbirds; the shape of the bill showing their al-
liance with the former, while the feet, wings, and
other prominent characteristics, establish their position with the latter.   The Meadow Lark and Hang-
ing-birds (incorrectly called Orioles) belong also to
this family, which differs very little from the Star-
ling group of the Old World.

The Cow-bird is one of those curiosities of Nature
for whose singular habits it is difficult to account.
Like the Cuckoo of Europe, the female builds no
nest of her own, but confides the care of her young
to various small birds, by watching their absence

from home, and then quietly dropping an egg into their nests. Half-a-dozen of these eggs may be found sometimes in a single morning's walk, by examining the newly made nests of the Yellow-poll Warbler, White-eyed Vireo, Blue-grey Fly-catcher, Golden-crowned Thrush, and Maryland Yellow-throat; the latter more especially seeming to be the favorite recipient of this unwelcome gift. It is a fact worthy of remark, that, although the Cow-bird is much larger than most of these birds, yet its egg is quite small, and approaches very near to the size of those in the nest where it is laid; it also, in almost every case, is hatched several days in advance of the others, thus securing to the young Cow-bird the exclusive care of its foster-mother. Her own eggs soon becoming worthless for want of attention, are tossed from the nest to make room for the fast-growing intruder, toward which she is as devoted in her attentions as though it was her own progeny.

As these birds do not pair, their life must neces-sarily be very different from that of others. While all around them are in the settled enjoyment of their mated companionship, the Cow-birds are roaming about the country in small companies, mingling pro-miscuously with each other, and seeming to have no particular preference for any stated locality. Early in the autumn the young birds instinctively join the old ones, when they assemble in flocks of immense size, and may be seen by the thousands and tens of thousands among the reeds along the river banks, and in the salt marshes.* After this they take their departure for the south in company with the Red-winged Blackbirds, assisting them in their autumnal depredations among the corn and rice.

There are some objects in the Creation whose utility we are sometimes inclined to question. How often, for instance, do we hear people wondering what mosquitoes were ever made for. It is true they are troublesome little pests, but they undoubtedly have their use, whether that use has yet been discovered or not. Thus it was for many years with the poor despised and hated Red-winged Blackbirds, which were looked upon by our farmers as little short of a scourge. Means of various kinds were devised to prevent their approach, but to little or no purpose, and the entire extermination of the race was looked upon as the only remedy for the evil; consequently the havoc which the murderous gun made upon their ranks was great. But how is it now? It has been observed that the amount of good they do silently in the spring more than compensates for the mischief they do in the autumn. If a flock of birds alights upon a field of standing corn, the inference is that they have come to steal; while if the same flock should settle upon a piece of fresh-ploughed ground where there is no crop to suffer from their depredations, but little notice is taken of it, when perhaps they may be rendering us signal service. So for years the poor Red-wings have suffered from the un-

---

* Letter from Dr. T. M. Brewer, of Boston, to J. J. Audubon.

4

just conclusions which we had drawn in reference to
their real merits.

Every farmer knows that fresh spring ploughing
turns up an army of grubs, worms, and the larvæ of
myriads of insects, which, if left to themselves, would
be sufficient to destroy a large portion of the crop
which the ground would produce. But just at this
time come the immense flocks of Red-wings and
Purple Grakles, which have been equally objects of
the farmer's aversion, and as they subsist almost exclusively upon this kind of food, they resort at once
to the open fields and cultivated grounds, where they
fully compensate the farmer for the few ears of corn
which they destroy in the autumn.

Red-winged Blackbird.

The Red-winged Blackbird generally selects for
a breeding place a low marshy piece of ground, occasionally interspersed with clumps of alder and
other bushes, among which or in a tall tussuck of

grass he builds his nest, composed of a mass of dry weeds or some other material for an exterior, and lined with fine grass or horse-hair. The female lays from four to six eggs of a light blue color, slightly spotted with brown. It is after the second brood is fully fledged that these birds congregate in such vast numbers, and commence their depredations upon the growing corn, which, being still young and tender, attracts them in such numbers as to darken the air and fairly to blacken the spot upon which they settle. At such times scare-crows avail little to protect the grain, and even the report of a gun will but drive them from one part of the field to another. This, however, does not continue long; as the corn advances toward maturity, it soon becomes too hard for their tastes, and away they fly to try their chance among the rice-fields of the South. The plumage of the male bird is very beautiful, and one of these vast flocks in their early spring dress presents a very grand and imposing appearance; their bodies of jetty black, with a broad patch of bright vermilion on each shoulder, which sparkles in the sun's rays with pleasing effect.

It is very interesting, in studying the habits of birds, to notice the peculiar methods adopted by the different species in the construction of their nests. While we see that some, with careless ease, build a fragile tenement upon the ground, others, with the skill of an architect, put together a most elaborate structure, such as would almost defy art to imitate. The bird we are now about to introduce is in this respect a very expert mechanic, weaving for itself a beautiful pensile nest of the most delicate texture, more or less perfectly wrought according to the skill of the workman, as it is sometimes observed that the older birds are the best builders.

Baltimore Oriole.

The Baltimore Oriole, or Hanging-bird, as it is sometimes called, is one of the most prominent of our summer visitors, both for its brilliant coloring and its lively and cheering song. Rare are the farm-houses where its black and orange-colored plumage may not be seen as it moves among the surrounding foliage, and where its voice is not heard welcoming the dawn. Its notes are few and simple, but their peculiar sweetness and harmony cannot fail to charm the ear.

In Louisiana, where the climate is often extremely warm, the Oriole suspends its nest from the north side of the tree, where it will be most sheltered from the sun, using in its construction the long fibres of the Spanish moss which it attaches at both ends to the forks in a branch, forming a number of loops about seven inches in length. When a sufficient number of these loops are made, it commences weaving in an opposite direction with the same material until it has produced a strong but open and airy pouch or bag, rounded at the bottom, and larger than at the top, where an aperture is left just large enough to admit of the easy passage of the birds in and out. There is no lining to this nest generally, as it is not required for warmth. In New York and Pennsylvania, where the atmosphere is cooler, and where there is a frequent occurrence of cold rains, it selects warmer materials, such as cotton yarns, hemp, tow, hair, wool, pieces of twine, or strings of any kind; these it uses in the same manner as its southern neighbor, but the texture of the nest, when complete, is firmer, more compact, and is furnished with a warm lining of cow's hair or wool. It is generally suspended upon the south side of the tree, where, while it can be well protected by the overhanging leaves from drenching rains, it is still open to the rays of the sun. The long, pendent boughs of the willow are a favorite resort of the Oriole, and here the female may be seen sitting quietly and at her ease with her nest flying in the wind in the midst of a violent storm; but so firmly is her house secured, that unless the branch from which it hangs should be torn from the tree, she need fear no harm.

There are two other species of Hanging-bird or
4*

American Orioles in the United States. Bullock's Oriole, which enjoys a wide range upon the Pacific coast, from California northward to the Columbia river, seems to fill the same position as that occupied to the eastward by the Baltimore Oriole, which it very much resembles in appearance as well as in its habits.

The Orchard Oriole is a familiar occupant of our orchards and gardens in summer, where it renders signal service by ridding the fruit trees of hosts of worms and noxious insects and their larvæ. It also suspends its neatly formed nest from the forks of some outspreading branch. It is not built in the pouch-like form we have before described, but looks more like a suspended cup, of insufficient capacity to conceal the body of the bird while sitting. It is a plainly colored bird; in the male the breast and whole lower parts, together with the rump, being of a rich chestnut-brown, and the remainder of the plumage black. The female is plain olive on the upper parts, and a dingy yellow below.

In point of nest-building we will now notice a bird of very different character; this is the Meadow Lark, a plain and humble species, seldom indulging in any wandering desires, not being gifted with any great powers of flight; its body being heavy and its wings short, and altogether unfitted for rapid motion. When it first rises from the ground, it flutters like a young bird until it rises fifteen or twenty feet in the air, when it pursues a bee-line course, with alternate sailings, and flutters until ready to alight, which is not often at any great distance, except during its migrations. Its nest is a loose structure composed of grass, fibrous roots, etc., and is placed at the base of a tuft of weeds or grass, in a small cavity scooped

Meadow Lark.

out of the earth; it is partially concealed from view by being covered with leaves and by the blades of growing grass drawn around it. The Meadow Lark justly merits a prominent place among our song birds for the sweetness and plaintive melody of its few simple notes, with which, in company with the Wood Thrush, it is among the first to welcome the dawn. The male and female are quite similar in their appearance, being mottled with brown and fawn color upon the head, back, and wings, while the chin and

breast are bright yellow; the throat being crossed
with a broad crescent-shaped band of velvety
black.

We must now leave these " creatures of music and
of song," and listen for a while to the cawings of a
ruder class, such as the Raven, Crow, Magpie, and
Jay, among which, however, we shall find much to
interest and instruct us.

Of the Crow family the Raven is the most prominent on account of its size, as well as its many singular qualities. From very early ages it has been
regarded with reverence and awe by the superstitious,
as being possessed of something unearthly in its

nature; in heathen
countries, especially,
it has been looked
upon by both priests
and people as a fore-
teller of events. In
some of the Indian
tribes of North Ame-
rica, their priests
wear, as a mark of
their sacred profes-
sion, two or three Ra-
ven skins affixed to
the girdle behind the
back, in such a man-
ner that the tails
stick out horizontal-

Raven.

ly from the body. They have also a split Raven skin

on the head, so fastened as to let the beak project from the forehead.*

The American Raven is a scarce bird in some districts, it being seldom seen, and consequently its character but little known. The European species is more abundant, and is often a very familiar bird. They are said to live to a great age, and the same pair have been known to resort to one spot to build for many successive years.

It is remarkable for having been the first living creature that left the ark after the flood; and as an instance of the great powers of wing of which it is possessed, we read that while the Dove which was afterward released could find no rest for the sole of her foot, and returned again to the ark, the Raven went to and fro upon the face of the earth until the waters were dried up.

Young Ravens may be tamed so as to become very amusing pets, but require almost constant watching, as they are mischievous and greatly addicted to thieving. A gentleman's butler having missed a number of silver spoons and other articles, without a suspicion as to who might be the thief, at last discovered a tame Raven with one in his mouth, and after following him to his hiding-place, found more than a dozen.† They are, however, gifted with some good qualities, being often possessed with a marked affection toward other animals, and also toward those with whom they have become familiar. A curious instance of attachment in a Raven is related as having occurred

* Stanley's "Familiar History of Birds."   † Ibid.

some years ago, at the Red Lion Inn, Hungerford,
England. A gentleman who lodged there coming
into the yard with his chaise, accidentally ran over
and bruised the leg of a favorite Newfoundland dog,
and while the injury was being examined, Ralph, the
Raven, looked on also, and was evidently making his
remarks on what was doing; for the minute the dog
was tied up under the manger with the horse, Ralph
not only visited him, but brought him bones and
showed him many other attentions. The gentleman
making some remarks to the ostler on the subject, he
was informed that the bird had been brought up with
a dog, and that the affection between them was mu-
tual, and all the neighborhood had been witnesses to
the many acts of kindness performed the one to the
other. The dog in course of time had the misfortune
to break his leg, and during the long period of his
confinement the Raven waited on him constantly,
carried him his provisions, and scarcely ever left him
alone. One night, by accident, the stable door had
been shut, and Ralph had been deprived of his
friend's company all night; but the ostler found, in
the morning, the door so picked away, that had it
not been opened, in another hour Ralph would have
made his own entrance.*

We will not say that it was because of this natural propensity of the Raven to form close and warm
attachments, that it was chosen by the Almighty to
carry food to the Prophet Elijah, during his solitary

---

* Stanley's " Familiar History of Birds."

sojourn by the brook Cherith, but the coincidence is certainly curious and interesting.

" In the United States the Raven is in some measure a migratory bird, individuals retiring to the extreme south during severe winters, but returning toward the Middle, Western, and Northern Districts at the first indications of milder weather. A few are known to breed in the mountainous portions of South Carolina, but instances of this kind are rare, and are occasioned merely by the security afforded by inaccessible precipices, in which they may rear their young. Their usual places of resort are the mountains, the abrupt banks of rivers, the rocky shores of lakes, and the cliffs of thinly peopled or deserted islands. It is in such places that these birds must be watched and examined, before one can judge of their natural habits, as manifested amid their freedom from the dread of their most dangerous enemy, the lord of creation.

" The flight of the Raven is powerful, even, and at certain periods greatly protracted. During calm and fair weather it often ascends to an immense height, sailing there for hours at a time; and although it cannot be called swift, it propels itself with sufficient power to enable it to contend with different species of Hawks, and even with Eagles when attacked by them. It manages to guide its course through the thickest fogs of the countries of the north, and is able to travel over immense tracts of land or water without rest." *

---

* Audubon.

CHAPTER III.

INSESSORES: *PASSERES—OSCINES.*

THE CROW — COMMON AND YELLOW-BILLED MAGPIE — BLUE
JAY — THE CANADA, FLORIDA, ULTAMARINE, STELLER'S,
MEXICAN, AND PRINCE MAXIMILIAN JAY—GREAT AMERICAN
SHRIKE — SOLITARY, WHITE-EYED, AND YELLOW-THROAT
VIREO—CEDAR-BIRD—WHITE-BREASTED NUTHATCH.

OF all the feathered inhabitants of America, with
which we are acquainted, the Crow is probably the
least of a favorite.   Having little either in his appearance or habits to recommend him, he seems to
be regarded by general consent as a plundering vaga-
bond, toward whom neither indulgence nor mercy are
to be extended; and were it not that a beneficent
Providence has gifted him with more than common
sagacity, the race, in our agricultural districts at least,
would have long since suffered a considerable dimi-
nution of numbers.   Watch the motions of yonder
sportsman with his double-barrelled gun, as he cautiously follows the windings of that old worm-fence,
upon a distant stake of which are perched two or
three ominous-looking birds, while a dozen or more
of the same sort are quietly rooting up the fresh-
sprouted corn in an adjoining field.   Well aware that
the watchful eye of the sentinel is ever on the look-
out for the approach of an enemy, he moves stealthily

along, until, fearful of losing his chance, he aims the piece at the nearest bird, who, immediately perceiving his danger, utters the alarm-note, and the whole flock follow his lead beyond the reach of powder. Sometimes the sportsmen conceal themselves among shrubbery in the track of the Crows as they pass to and from their roosting-places, but even here they cannot always escape the scrutinizing glance of these ever-suspicious birds, for they may be observed to wheel to the right or the left of the spot as soon as they approach within a short distance of it. A constant fear of falling a prey to the murderous gun seems to attend the whole life of the poor Crow; every suspicious-looking individual is avoided with care, and it is almost impossible to come within shooting distance of him without great caution.

But why is it that this bird should thus be an object of common hatred and execration? Simply because, as in the case of the Blackbirds, we have placed a wrong estimate upon the works of an All-wise Creator. What if the Crow does root up the corn in some places, compelling the farmer to replant and replant until his patience is gone? The cutworms, of which these injured birds annually destroy myriads, are certainly a far worse enemy, and more to be dreaded, inasmuch as they appear when the crop is far advanced, and accomplish its destruction when it is too late to replant.

The nest of the Crow is generally built in some quiet and secluded spot, upon the jutting crag of a precipitous rock, or among the thick branches of

some tall tree beyond the reach of his great enemy—
man.   It is composed externally of moss, sticks, and
thin pieces of bark, stuck together sometimes with
mud or clay, and lined with horse-hair or wool, so as
to make a thick warm bed.   The eggs are four, of a
pale greenish hue, marked with blotches of olive.
When the vicinity of the nest is approached, the
noise made by the birds often brings to their assistance all the Crows in the neighborhood, who join in
the general hubbub until the intruder retires, frequently following him to a considerable distance, as
though to be sure of his retreat.

In the Autumn these birds congregate in vast
flocks, and resort to some particular spots to roost,
generally along the margins of rivers or the shores
of lakes, where there is an abundant growth of reeds,
upon which they settle in such numbers as to bend
them to the earth.   Toward these roosting-places
they may be seen, in the latter part of the day,
slowly wending their way, in long, straggling, and
apparently interminable lines, sometimes flying low
over the fields, and sometimes high above in the air.
These flocks disperse during the daytime in smaller
companies to search for food.

The Crow is capable of being domesticated so as
to become quite an amusing pet, and, it is said, may
be taught to utter a few words of good English.   It
soon learns to distinguish between the different members of the family, appears terrified at the approach
of a stranger, has a great propensity for hiding small
articles, particularly of metal, also corn and food

generally. He is fond of the company of his master, and will recognize him even after a long absence, as the following well-authenticated anecdote will show :

"A worthy gentleman who resided on the river Delaware near Easton, had raised a Crow with whose tricks and society he used frequently to amuse himself. The Crow lived long in the family, but at length disappeared, having, as was then supposed, been shot by some vagrant gunner, or destroyed by accident. About eleven months after this, as the gentleman, one morning, in company with several others, was standing on the river shore, a number of Crows happening to pass by, one of them left the flock, and flying directly toward the company, alighted on the gentleman's shoulder, and began to gabble away with great volubility, as one long absent friend, naturally enough, does on meeting with another. On recovering from his surprise, the gentleman instantly recognized his old acquaintance, and endeavored, by several civil but sly manœuvres, to lay hold of him; but the Crow, not altogether relishing quite so much familiarity, having now had a taste of the sweets of liberty, cautiously eluded all his attempts; and suddenly glancing his eye on his distant companions, mounted in the air after them, soon overtook and mingled with them, and was never afterward seen to return."*

The Magpie, which in Great Britain is so common

---

* Wilson's "American Ornithology."

and familiar a bird, is comparatively little known in the United States, its haunts being strictly confined

Magpie.

to the vast territory lying west of the Mississippi, where, in some districts, it appears to be abundant. It is represented as a bold, active, and restless species, constantly flying from place to place; being possessed of all the voracity peculiar to his tribe, and very fond of the eggs and young of other birds, especially Chickens, Pheasants, and Partridges; it will even alight upon the backs of cattle, and peck out the larvæ of insects that are secreted in the skin, and is quite well satisfied with carrion if no better food is at hand.

" In 1804, an exploring party, under the command of Captains Lewis and Clark, on their route to the Pacific Ocean, across the continent, first met with the Magpie somewhere near the great bend of the Missouri, and found that the number of these birds increased as they advanced. Here also the Blue Jay disappeared; as if the territorial boundaries and jurisdiction of these two noisy and voracious families

of the same tribe had been mutually agreed upon
and distinctly settled. But the Magpie was found
to be far more daring than the Jay, dashing into
their very tents, and carrying off the meat from the
dishes. One of the hunters who accompanied the
expedition stated that they frequently attended him
while he was engaged in skinning and cleaning the
carcass of the deer, bear, or buffalo he had killed,
often seizing the meat that hung within a foot or two
of his head. On the shores of the Koos-koos-ke
river, on the west side of the great range of Rocky
Mountains, they were found to be equally numerous.

"It is highly probable that those vast plains or
prairies, abounding with game and cattle, frequently
killed for the mere hides, tallow, or even marrow-
bones, may be one great inducement for the residency
of these birds, so fond of flesh and carrion. Even
the rigorous severity of winter in the high regions
along the head-waters of the Rio del Norte, Arkansas,
and Red Rivers, seems insufficient to force them from
those favorite haunts; though it appears to increase
their natural voracity to a very uncommon degree.
Colonel Pike relates that in the month of December,
in the neighborhood of the North Mountain, these
birds were seen in great numbers. 'Our horses,'
says he, 'were obliged to scrape the snow away to
obtain their miserable pittance; and, to increase their
misfortunes, the poor animals were attacked by the
Magpies, who, attracted by the scent of their sore
backs, alighted on them, and in defiance of their
wincing and kicking, picked many places quite raw;

5 *

the difficulty of procuring food rendering these birds so bold, as to alight on our men's arms, and eat meat out of their hands.' " *

There are two species of Magpie found within the limits of the United States; the Common Magpie, which we have just described, and the Yellow-billed Magpie, both of which may be styled showy and ornamental birds. Their long, wedge-shaped tails, composed of beautifully colored feathers of brilliant blue and shining green, give them a peculiarly elegant and graceful appearance. The head, neck, back, and throat of the Common Magpie are black, the lower parts, together with the scapulars, white, the tail, upper wing-coverts, and secondary quills of the wings, are rich green with purplish reflections. The Yellow-billed Magpie is very similar to the above in size and appearance, except that the bill is bright yellow, and the crown of the head is glossed with green. It is a resident of Upper California.

The family of birds of which the Common Blue Jay is the principal representative in the United States, probably enjoys as wide-spread a reputation as any other division of our American Fauna. It is said that with the exception of Southern Africa, Australia, and the Pacific Islands, there is no country upon the globe where some of its representatives are not found. But it is on the American Continent that it is most abundantly diffused, especially in Mexico and the countries lying adjacent to the Equator, where there are many beautiful species, displaying a great

* Wilson's "American Ornithology."

variety of the most exquisite tints in their plumage. A prevailing color of the whole group is blue, of different shades, from a light azure or ultamarine, to a deep, dull indigo. In the United States we number about eleven species. In the north and east we have the Blue Jay and the Canada Jay; in the south, the Florida Jay; and in the west and north-west, the Ultamarine Jay, Steller's Jay, Prince Maximilian's Jay, Mexican Jay, and Beechy's Jay, the two latter being mostly confined to Texas and California.

Nearly all our country boys are familiar with the Blue Jay, with his high, peaked crest, his black

Blue Jay.

whiskers, and his broad wings and tail so beautifully banded with blue, black, and white. His bold, sprightly bearing, his malicious and deceitful habits, his sly and cunning disposition, and his great fondness for tasting the eggs which other birds have laid, are facts in his history well known to most. His

showy plumage, attractive form, and graceful motions, as well as his restless and noisy activity, render him one of the most prominent inhabitants of our
woodlands.   It is difficult to realize how a creature
so eminently favored and gifted with so many per-
sonal charms, should also be possessed of so much
selfishness, mischief, and malice.   But so it is ; and
even in the beautiful garb of the Blue Jay we find
the moral written, that it is unsafe to judge from ex-
ternal appearances.   It robs the nests of other birds
indiscriminately, sucking the eggs or devouring the
young, and will even attack large birds and other
animals which have been wounded or otherwise dis-
abled ; but true to his cowardly disposition, he seldom risks his safety in open combat with his equals.
Audubon says, "The Cardinal Grosbeak will chal-
lenge him, and beat him off the ground.   The Red
Thrush, the Mocking Bird, and many others, although
inferior in strength, never allow him to approach
their nests with impunity; and the Jay, to be even
with them, creeps silently to it in their absence, and
devours their eggs and young whenever he finds an
opportunity.   I have," he adds, "seen one go its
round from one nest to another every day, and suck
the newly laid eggs of the different birds in the
neighborhood, with as much regularity and composure as a physician would call on his patients.   I
have also witnessed the sad disappointment it experienced, when, on returning to its own home, it found
its mate in the jaws of a snake, the nest upset, and
the eggs all gone."

The Canada Jay is a very plain and unpretending bird, being the only species of those we have named which is destitute of a brilliant plumage; its predominating colors being a dull slate and drab, with occasional markings of black. It inhabits the State of Maine, Nova Scotia, and New Brunswick, and in winter a few individuals are seen as far southward as Pennsylvania. It is abundant in the Canadas and Labrador, and has been found in the vicinity of Fort Astoria, on the Columbia river. It becomes very familiar with the wood-cutters of Maine, entering their camps on very social terms, and helping itself to such pieces of flesh as are within its reach. These wood-cutters sometimes "amuse themselves in their camp during their eating hours, with what they call 'transporting the carrion bird.' This is done by cutting a pole eight or ten feet in length, and balancing it on the sill of their hut, the end outside the entrance being baited with a piece of flesh of any kind. Immediately on seeing the tempting morsel, the Jays alight on it, and while they are busily engaged in devouring it, a wood-cutter gives a smart blow to the end of the pole within the hut, which seldom fails to drive the birds high in the air, and not unfrequently kills them." *

Prince Maximilian's Jay was first discovered in the Rocky Mountains by the celebrated naturalist whose name it bears, while travelling in the interior of North America. In form and general appearance, as well as in other important particulars, it differs

* Audubon.

from any other member of the group. The absence of the long expansive tail, which adds so much grace to the motions of other species, and its peculiarly short, clumsy figure, are very apparent. It, however, possesses in a very high degree the carnivorous and rapacious propensities of its tribe, living mostly upon frogs, lizards, and other reptiles. It appears to be rather a scarce bird, and quite difficult to procure on account of its shyness. Specimens are therefore seldom seen in our cabinets.

The Shrikes in many respects closely resemble the Jays. With the exception of the head and bill, in general form they are not unlike; in manners and habits they still more closely agree, and although some naturalists have assigned them a place immediately following the Hawks, in consequence of the shape of the head, being broad and stout, and also the hooked form of the bill, yet the other characteristics by which the genera are determined are unquestionably in favor of their position being near the Crow family.

The Great American Shrike, or Butcher Bird, is more than a match for the Blue Jay in cruel rapacity. Its food consisting almost exclusively of large insects, birds, and the smaller quadrupeds, it has attained the reputation of being an expert hunter, seizing upon its prey with great dexterity, much after the manner of the Sparrow Hawk. It is said to possess the faculty of imitating the notes of other birds, especially such as are indicative of distress, which it does no doubt for the purpose of decoying them

within its reach, as it has been known upon such occasions to dart suddenly into the thicket and bear off
the body of some deluded victim. It will also occasionally pursue its prey upon the wing for a considerable distance, and
sometimes succeeds in
bringing it to the
ground.

Like the Jays, this
bird has the habit of
stowing away its sur-
plus food, as though
for future use. The
Jay finds some hole
in a tree, or crack or
crevice in the bark,
where he secretes
what he does not need
for the present; while
the Butcher Bird im-

Butcher Bird.

pales its victims upon thorns or other sharp points
that may happen to suit its purpose. The object of
its so doing remains a mystery, many opinions relative to the subject having been expressed by various
observers, but this part of the history of this singular
bird has yet to be properly elucidated.

The Butcher Bird inhabits most of the Eastern,
Middle, and Southern States, retiring during Summer to the more northern and mountainous districts,
for the purpose of incubation. The nest is generally
built among the forked branches near the top of a

small tree; it is composed externally of grass, moss, and leaves; internally of fibrous roots, and is warmly lined with feathers. The eggs are mostly four in number, of a dull ashy hue, spotted and streaked toward the great end with brown.

Among the many charms which attend upon the opening of Spring, and by their sweetness and beauty do much to render its advent a joyous and lovely season, one of the most pleasing is the song of the birds. How gentle and soothing are their tones, as, with the highest glee, they warble out their inspired music! How peaceful are the thoughts that occupy the mind that has been abstracted from itself, by the distant voice of some modest little bird! We can hardly place too high an estimate upon the kindly influences which they exert upon us. Listen to the persuasive tones of the Red-eyed Vireo, — soft and sweet, and full of eloquence, bidding us cast aside our griefs, and be as happy as he. And the Warbling Vireo— light-hearted little fellow, he too tells us that the skies are bright and the sun is ever shining, notwithstanding clouds may obstruct them from our view; and though the day be dark, he sings on, still looking for a bright to-morrow.

These two charming little songsters belong to a group formerly placed with the Fly-catchers and Tyrants, on account of their possessing some of the habits of those birds; but a slight comparison will at once show the difference. The true Tyrants are of a stout, heavy build, particularly about the head and shoulders, while the Vireos are light and delicate, approaching nearer the Warblers in that respect. The Vireos are more musical than the Tyrants, the latter being as a family quite destitute of song. Their motions upon the wing are also quite dissimilar; the flight of the former being gliding, and with little motion of the wings, while that of the latter is accompanied frequently by rapid fluttering. A position in the family of Shrikes is now believed to be the most natural. There are a number of species of the Vireos found in our woods, but the two mentioned above are the most conspicuous on account of the peculiar sweetness of their notes. Besides them we have the Solitary Vireo, the Yellow-throated Vireo, and the White-eyed Vireo; the latter has sometimes been called the Politician, in consequence of their nests being seldom found without containing one or more pieces of newspaper in their texture. The nests of all these little birds are particularly neat in their structure, being mostly composed of fine materials, and arranged with the utmost skill. That of the White-eye is built in a low bushy vine, a species of Smilax which is very abundant. It is in the form of an inverted cone, and besides the newspaper, we find small dry twigs, grasses, and pieces of hornet's nests; the whole is lined with fine fibrous roots. In this snug little cavity is often found the egg of the Cow-bird, several of the species of the Vireos being honored with the task of assisting to perpetuate this singular race.

The next bird we shall present is the Cedar-bird, or Waxwing. Arrayed in a plain and modest suit, possessed of no song, its only note being a low monotonous lisp, scarcely audible at a distance of fifty

paces. Yet to it is given a plumage of the most exquisitely soft and silky texture, which lays so close and smooth, that the webbings of each feather are scarcely distinguishable. The head is surmounted by an ornamental crest, capable of being raised or depressed at the will of the bird. The general color of the plumage is a beautiful fawn, lightest

Upper fig.—Red-eyed Vireo.
Lower fig.—Cedar-bird.

upon the lower parts; a band of velvety black margined above and below with white, passes from the forehead over the eye toward the hinder part of the head. The tail feathers are all broadly tipped with bright yellow; the shafts of each, together with those of the secondary quills in the wings, being extended in a short appendage very much resembling a small piece of red sealing-wax. Those upon the tail are seldom found perfect, as they are soon worn off among the thick foliage of the cedars.

The Cedar-birds are very abundant in spring and

fall, associating in flocks of considerable size, moving about in a compact body, and alighting in such numbers and so closely together, as almost to touch each other. Now is the sportsman's opportunity, as a dozen or more may be brought down at one shot, but so soon as they commence to disperse over the tree, which they do almost immediately, they become restless, and are more difficult to kill. At these seasons of the year their appetites are so voracious as to lead them to devour almost everything in the shape of fruit which comes in their way; hence they become very fat, and are considered as excellent eating, large numbers being exposed in our markets for sale.

It is a singular fact that the Cedar-bird, although one of the earliest of our visitors, is probably the last to commence the important business of incubation, thousands of young birds of other species having left their nests before it has begun to build. They seem also to have less regard for the safety of their young than most birds, as the nest is placed in a low horizontal branch of an apple-tree; and when approached the bird flies off without any manifestation of concern.

We should scarcely be doing our readers justice, were we to omit noticing another family of quiet little birds which are the farmer's special friends. Of these the White-breasted Nuthatch is probably the most familiar representative in the Middle States. It is a social and familiar species, frequenting the garden and orchard, and even the house-top, where

it may be seen sometimes picking in the decayed
shingles for insects.   As though aware of its useful-
ness, it confidingly trusts in the protection of man,
which indeed should always be afforded it, as its
busy, active life is almost constantly devoted to his
service, in the destruction of myriads of hurtful insects and their larvæ. Its note is a low chirp, which
it occasionally utters as it runs up and down the
trunks of the trees in search of food. It is extremely
nimble when so engaged, moving with great facility
sideways or head downwards, in which position it
will sometimes stop and gaze with a peculiarly quaint
expression at the bystander for some time, although
he may be within a few paces of the tree, and then
with a sudden jerk of its wings, off it goes to an adjoining tree.   During summer they retire to the
woods, where they dig out a small hole in some decayed tree with their long sharp bills, at the bottom
of which the female lays her eggs. There are three
other species of this family in the United States, all
of them active and pretty little creatures. The Brown-
headed Nuthatch is quite abundant in the South,
where it is a constant resident, but is seldom found
north of the State of Maryland; while the Red-bel-
lied Nuthatch occupies the eastern and northern districts; its occurrence south of Pennsylvania is rare.
The latter is quite as abundant in the mountainous
parts of Pennsylvania as the White-breasted species.
Of the California Nuthatch but little is known, but
as its name implies, it is an inhabitant of Upper California.

We cannot too forcibly impress upon our country friends the value they should attach to the lives of these little birds, and unless for scientific purposes they should positively prohibit their being destroyed. The idea that they suck the sap from the tree is too preposterous to be listened to. But the assertion that they save an enormous amount of fruit from destruction by insects, is true.

CHAPTER IV.

INSESSORES: *PASSERES—OSCINES.*

A STROLL INTO THE WOODS OF CAROLINA—MOCKING BIRD—
WOOD ROBIN — CAT-BIRD — ROBIN — GOLDEN-CROWNED
THRUSH — BROWN THRUSH — FINCHES — SNOW BUNTING —
SONG SPARROW—INDIGO BIRD—NONPAREIL—HOUSE FINCH
—CROSSBILL—CARDINAL GROSBEAK—BLUE GROSBEAK—
SCARLET TANAGER.

WE will now take a stroll into the woods of Carolina, and see if there is anything among the feathered
inhabitants of the South that is peculiarly attractive
and worthy of attention. The lofty branches of the
Long Leaved Pines are waving majestically over our
heads, their fine and beautiful foliage being here and
there varied by a clump of Oaks and Hickory. Clus-
:ering vines of many sorts are twining themselves
around the giant stems, wreathing the branches with
festoons of gay-colored flowers, and mingling their
fragrance with the sweet bloom of many a flowering
shrub. The air is soft and balmy, and possessed of
a peculiar freshness which is characteristic of a pine
forest. Nature here seems to have profusely spread
her charms on every side, pointing us at each step to
some new object of admiration. The mellow whistle
of the Red Bird is heard overhead, together with the
call of the Jay, the soft warbling of the Vireos, the

Upper fig.—Blue Grosbeak.   *Lower fig.*—Scarlet Tanager.

(67)

mewing of the Cat-bird, the loud clear melody of the Wood Robin, the shrill cry of the Woodpecker, and many other voices equally attractive. As we advance into the thicket the confusion of sounds increases; every song with which we are familiar, and many more, seem suddenly let loose upon the ear, and last, though it is broad day, we are surprised to hear the cry of the Whip-poor-will. This exciting our curiosity, we naturally look around to discover the cause of so strange an occurrence; but still sounds the clear cry of "Whip-poor-will! whip-poor-will!" When suddenly there darts up from a low bush near by a fine Mocking Bird, and settling on a branch within our view, continues his varied melody. The secret is at once explained; the little mimic before us has been the sole cause of our surprise, and there he sits flirting his long tail from side to side with an air of perfect satisfaction, expanding his wings and stretching his neck in all directions, while he calls out with much animation, "Bob-White! Bob-White!" and before poor Bob-White has time to scamper to his covey, the screams of the Pigeon Hawk are heard wild and clear; then immediately his voice falls into some soft and tender warble, gradually rising higher and higher until we recognize, among a host of others, the clear and ringing melody of the Brown Thrush, set off with the gentler tones of the Robin and Blue-bird, occasionally interspersed with some fine, pleasing original notes. We stand and listen with delight to this grand concert of Nature's great musician, his voice ever changing, ever sweet, until the twilight unconsciously steals upon us; still the serenade continues. The pale moon glimmers in the eastern sky, and as it grows brighter and brighter and darts its vivid beams into the forest's deep recess, our little performer, as though animated with fresh spirit, seems to strain his utmost powers in pouring forth a flood of the most enchanting song. This exquisite aërial music is often maintained during most of the night, or until the moon sets, two or three birds sometimes vying with each other in the strength of their voices.

Mocking Bird.

But it is not in the Southern States alone that the Mocking Bird is found, it having been seen in Pennsylvania and even as far north as Boston; in these places it is generally shy, and does not sing with that energy which characterizes it at the South. In Louisiana it remains as a constant resident, feeding during winter on the numerous berries and small fruits which are mostly abundant.

In the Northern and Middle States, the place of the Mocking Bird is filled by the Wood Robin or Wood Thrush, a bird whose song, although not so varied as that of his southern neighbor, is nevertheless, on account of its peculiar power and sweetness, not less pleasing. Audubon says that among all the feathered tribes of the woods, this is his favorite; and we question if this is not the general opinion of most of those who are accustomed to the different notes of our best songsters.

It delights in deep shady woodlands, where there is a thick undergrowth of low shrubbery, and through which meanders some shallow brook, that

> "Sparkles out among the ferns,
> To bicker down a vale."

Here its loud, clear, and mellow voice may be heard almost from morning till night, but more particularly just after daybreak, and in the evening, continuing it until some time after sunset.

There are various kinds of Thrushes which are interesting, but which our limits will hardly admit us fully to describe. With the Robin almost every child that has been much in the woods is familiar. It is gifted with considerable powers of song, the note being a loud, clear, and very musical warble, uttered with much force and rapidity. We have also the Golden-crowned Thrush, which builds a nest

Robin.

upon the ground so resembling an oven as to have given it with some the title of Oven Bird. It also is very abundant in the woods during summer, and has a loud, shrill voice, which can hardly be called musical.

Another favorite member of this family is the Brown Thrush or Thrasher. Its song is very clear and melodious, resembling more that of the Wood Thrush, but not attended with that peculiar softness of expression which renders the latter a songster of superior merit.

The Thrushes appear to be gifted with considerable intelligence, and display much sagacity in protecting themselves and their young from danger. The Mocking Bird is often annoyed by the predatory visits of the Rattlesnake, against which he battles with great ferocity, avoiding dexterously his venomous bite, and at the same time pecking at his eyes and head with such vehemence as to drive him quite away from the spot. The Brown Thrush also finds an enemy in the Blacksnake, which quietly and almost unobserved crawls into the nest for the purpose of making a meal of the eggs or young. But no sooner do the old birds catch sight of him, than a general uproar ensues; other thrushes in the vicinity assemble at the cry of distress and boldly assault the intruder, fairly pecking the eyes from his head, and it may be well for the poor snake if he escapes without being stretched lifeless upon the ground.

An anecdote is told of a Thrush, of what species we are uncertain, who had built her nest near the spot where some rocks were being blown with powder. At first, whenever the blast would explode she was disturbed by the fragments of rock flying in all directions, but still she would not leave the spot. At length observing that just before the explosion there was a bell rung, upon which the miners immediately withdrew to a place of safety, she concluded to follow their example; accordingly, when the bell rung she retired to the spot where the workmen were sheltered, and dropped close to their feet, remaining until the explosion had taken place, when she returned to her nest. The story of her sagacity was soon told, and visitors wished to gratify themselves by observing the bird. But as explosions could not be produced whenever they pleased, the bell was rung instead, which for a time answered the purpose; but the bird was not to be so trifled with; she refused to leave her eggs merely to amuse her visitors,

7

and so when the bell was rung she peeped out to see if the workmen retreated, and if not, she remained quietly on her nest.

One of the sweetest as well as the most familiar notes with which we are acquainted, is that of the Bluebird. He is among the earliest visitors from the South, even coming to us from a great distance to pass a few warm, bright days before the close of Winter, disappearing, however, at the return of severe cold. But no sooner has the first breath of Spring offered him a more certain inducement to remain, than he is seen flitting cheerily about the farm-house and along the fence-rows, uttering his soft and plaintive warble with a degree of innocence which no sensitive heart could fail to appreciate. He early visits his old haunts about the wood-shed and out-houses, examining the spot where his last year's nest was built, and with all the ardor and zeal of a new-born affection he assists his mate in rearranging the materials for their abode, which is often in a box made for his use and nailed to a post in the garden; but not unfrequently he builds in the hole of some decayed tree or old gate-post. The writer once saw one of these nests which had been built at the bottom of a hole in a gate-post, from which it required some ingenuity on the part of the old birds to effect the escape of their young, the hole being too deep for them to get out alone. This difficulty they had overcome by placing a few small sticks on one side of the hole in the form of a ladder, by which means they could crawl out. The Bluebird sometimes no

Upper fig.—Hooded Warbler. *Lower fig.*—Bluebird.

(75)

sooner becomes nicely and to all appearance perma-
nently fixed in his snug little box, than he is attacked
with such vigor and determination by the Wren, that
he is compelled to give up the premises which he had
preoccupied, the latter not considering his more tardy
habits as in any way lessening his right to its occupancy, provided he can gain possession.

There is something sweetly attractive in the manners and habits as well as the song of the Bluebird.
Attaching himself closely to the habitations of man,
he seems to have become a sort of domesticated pet,
whose annual reappearance among us is welcomed
with peculiar pleasure. It is probable that he remains with us during a greater portion of the year
than any other migratory song-bird, unless it be the
Robin. Before the cold breath of Winter has passed
away, he comes to us fresh from a land of sunshine
and flowers, with a merry little heart beating high
beneath his bright chestnut bosom, and his little
throat seeming to stretch to its utmost capacity to
cheer our lone hours with his song. During Summer he is our ever-constant and ever-welcome companion. He attends us in our rambles, flitting before
us as we walk by the roadside. If we pass near the
place where his nest is built, he will mount a stake
near by and warble out his sweet little ditty, his
bright azure coat sparkling in the sun, as he nimbly
twits himself about, opening and shutting his wings
frequently and in rapid succession. He watches
carefully over the interests of the garden, and many
a noxious caterpillar is transferred from his lurking-

7 *

place among the vegetables to the mouths of his hungry little ones.

> " When all the gay scenes of the summer are o'er,
>   And autumn slow enters, so silent and sallow,
> And millions of warblers, that charmed us before,
>   Have fled in the train of the sun-seeking Swallow,
> The Bluebird, forsaken, yet true to his home,
>   Still lingers, and looks for a milder to-morrow,
> Till, forced by the horrors of winter to roam,
>   He sings his adieu in a lone note of sorrow."

There are two other species of Bluebirds in the United States, both of which are inhabitants of the far West. The Western Bluebird resembles our own closely, but is readily distinguished from it; the principal difference is that the chestnut of the throat extends in a collar round the neck. The Arctic Bluebird is a beautiful creature, the whole plumage being of a soft ultamarine, least brilliant on the throat and breast. It is found as far north as the mouth of the Columbia river.

Perhaps there is no family of the Passerine Insessores more numerous, or containing a greater variety, than the Fringillinæ or Finches. It consists of about nineteen subdivisions and comprises between sixty and seventy species, all inhabiting the territory of the United States. It would be impossible for us here to enter upon any very general description of these birds; we must therefore confine ourselves to a few of the most prominent individuals among them, leaving it to our readers to pursue the study of this interesting group as their inclination may lead them, by the assistance of more complete or voluminous works.

The cold and icy regions of Labrador and Green-land are inhabited by a number of beautiful birds, which are very rarely seen except during their short stay among us in winter, when the severity of the northern climate and the scarcity of food compel them to remove to a warmer latitude. Among these are the Lapland Longspur and the Snow Bunting. They are both quite abundant in the Western States during winter, but the former is seldom met with near the Atlantic coast, confining itself principally to the region of country lying north from Kentucky and west of Pennsylvania. In Nova Scotia and the States of Maine, New Hampshire, and Vermont, the Snow Buntings make their appearance in large flocks, about the time of the first fall of snow, and spread themselves over great tracts of country in search of grubs, larvæ of insects, seeds, and any other sub-stance that will answer the purpose of food; as the severity of the season advances, they proceed south-ward, occasionally, though rarely, as far as the vicin-ity of Philadelphia. In Summer these birds not only inhabit Labrador, Greenland, and Iceland, but even the piercing climate of Spitzbergen, where the in-tensity of the cold is such that vegetation is nearly extinct. Indeed, they seem to make the whole coun-try within the limits of the Arctic Circle, their home, from whence they spread themselves in vast numbers southward on both continents, upon the opening of the Arctic winter.

The plumage of the Snow Bunting varies so much with age, climate, and other circumstances, that it is almost impossible in the same flock to procure more

than a few specimens whose markings and colors are precisely similar. They are for the most part white, with various inter-mixtures of fawn color and black upon the head, back, shoulders, and wings. Some specimens are pure white, others white and black only, while in some the fawn color predominates.

Snow Bunting.

We must not confound the Snow Bunting with our familiar and welcome little winter visitant, the Snow Bird. Although both belong to the same family, they differ greatly in their size, appearance, and manners, the latter being fully one-third smaller. The predominating color of its plumage is a deep leaden brown, with white on the breast and sides, and two white feathers on each side of the tail. This little bird comes to us just as the ground is being strewn with the autumn leaves, and, continuing during the severest winters, leaves us again for the north early in the Spring. It is a sprightly and active as well as a social and confiding bird, entering the city in great numbers, so that there is scarcely an open garden where they may not be found picking up the crumbs and pieces of waste food that are thrown out. In the country they sometimes mingle in small flocks with the Tree Sparrows and Titmice. They seem particularly active just after a fresh fall of snow, flying about from bush to bush with apparent delight, twittering and chirping with great animation. We cannot help wondering what a winter would be without the Snow Birds; for however we may appear to be indifferent to their existence, they certainly exert a silent influence upon us, in enlivening and animating a scene which without them might be dreary and dull.

Among the many beautiful little Sparrows and Finches that tenant our groves during the summer months, the Song Sparrow and the Indigo Bird present themselves as objects peculiarly

Upper fig.—Snow Bird.
Lower fig.—Song Sparrow.

worthy of our attention. The former, although rather a plain and unobtrusive little fellow, still merits our

F

affection for the sweet and sprightly notes with which he cheers us so early in the Spring. Although partially migratory, yet in the warmer parts of Pennsylvania and New Jersey he may be considered as a permanent resident. Here his notes are heard in advance of the Bluebird. His song is not possessed of much variety, but is uttered with great force and sweetness. Most birds become quite silent after the brooding season is passed, but not so with our little Sparrow; he sings with as much animation in the Autumn as in the Spring; and sometimes even in the depth of Winter his clear and cadenced voice may be heard among the low bushes which skirt our woodlands.

The Indigo Bird, as its name implies, is gifted with a coat of the deepest and most brilliant blue. It is quite a small bird, about the size of the Chipping Sparrow, and in addition to its gay and attractive plumage, is possessed of a fine song. Mounted upon the top of a tall tree, it will sit for half an hour and chaunt its simple lay, which somewhat resembles that of the Canary, but is not so varied, commencing with a loud clear warble, and gradually falling for six or eight seconds until it is scarcely audible, and after a short pause repeating it without variation. Its favorite haunts are about gardens, clover-fields, the borders of woods, and the roadsides, where it may often be seen perched upon a fence-stake, singing with great vivacity. The female is not possessed of the same brilliant livery as her mate, neither do the males assume their perfect dress until the third season.

Closely allied to the Indigo Bird are the Lazuli Finch and the Painted Finch or Nonpareil. The former abounds in the western territories, from the Arkansas river to the Columbia, but is never seen to the eastward. The males are beautiful birds, and frequently indulge in a pleasing and not unmusical song. Their plumage is of a fine light blue, with a slight tinge of green, except on the breast and sides, which are white, intermingled with fawn. The Nonpareil is one of the most common and familiar birds in the Southern States, particularly in the lower part of Louisiana. In the vicinity of New Orleans they are so abundant in the Spring that almost every orange grove seems alive with them, and they may be seen flying along the roadsides in great numbers. When they first arrive from their winter quarters in Mexico, the males are very pugnacious and quarrelsome, and are almost continually engaged in fighting. This jealous disposition is made use of by the bird-fanciers to catch them alive in their traps, which they do in the following manner:

"A male bird in full plumage is shot and stuffed in a defensive attitude, and perched among some grass-seed, rice, or other food, on the same platform as the trap-cage. This is taken to the fields, or near the orangeries, and placed in so open a situation, that it would be difficult for a living bird of any species to fly over it without observing it. The trap is set. A male Painted Finch passes, perceives it, and dives towards the stuffed bird with all the anger which its little breast can contain. It alights on the edge of the trap for a moment, and throwing its body against the stuffed bird, brings down the trap and is made a prisoner. In this manner thousands of these birds are caught every spring." *

The beauty of the plumage of this little Finch, as well as the sweetness of its song, has rendered it a general favorite among those who are fond of keeping pets; it sings with great energy in confinement, and with care will live for eight or ten years.

Of all the gay-winged minstrels with which our country abounds, the Painted Finch is one of the most brilliantly attired. The head is of a beautiful cobalt blue, a patch of bright yellow covers the back and shoulders, while the rump and the whole lower parts, including the throat and breast, are of flaming scarlet. The females are plain greenish-olive above, and dusky yellow below; the young birds of both sexes assume this garb from the nest, the males gradually undergoing a change with each successive moult until about the fourth or fifth season, when their dress is complete.

In New Mexico and California there is a beautiful and familiar little bird called the American House Finch, which is probably as great a favorite among the people of those countries as the Barn Swallow, the Wren, and the Bluebird are with us. The following interesting description of its manners and

---

* Audubon.

habits, by Col. M'Call, we extract from Cassin's "Birds of America:"

"I found this charming little Finch abundant at Santa Fé (New Mexico), where it commenced nesting in March, although the weather was still wintry, and so continued, with frequent snow-storms, for more than a month. Notwithstanding this, the song of the male failed not to cheer his mate during incubation, with the liveliest melody. The notes often reminded me of the soft trill of the House Wren, and as often of the clear warble of the Canary. The males of the last year, though mated and apparently equally happy and quite as assiduous as their seniors, were not yet in full plumage, having little or nothing of the red colors that mark the adult birds.

"The nests which were stuck into every cranny about the eaves and porticoes of the houses throughout the town, were variously composed of dry grass, fine roots, horse-hair, long pieces of cotton twine, or strips of old calico; in fine, of countless odds and ends, that were picked up about the yards; — these were curiously and firmly interwoven, so as to make a warm and comfortable abode for the new-comers.

"His disposition toward other birds appeared to be mild and peaceful, as I had many opportunities to observe. I will mention one instance. In the piazza of the house I occupied, quite a colony of these birds had their nests: here the work of building and incubation had gone on prosperously for several weeks, although the weather at times was stormy and cold, and ere the genial warmth of Spring was fairly felt,

8

the colony might have been .said to be fully established. As the season advanced and birds of a
less hardy nature began to arrive from the South, a
pair of Barn Swallows made their appearance, and
forthwith entered the territory of the Finches. And
here they at once, very unceremoniously, began to
erect their domicil. This act of aggression would
have been fiercely resented by most birds, and violent measures would have been resorted to, to eject
the intruders. The conduct of the little Finches was
quite different; at first they stood aloof, and seemed
to regard the strangers with suspicion and distrust,
rather than enmity. In the meantime the Swallows
went quietly to work, without showing any inclination to intermeddle; and in a day or two (their mud
walls all the time rapidly advancing) they gained the
confidence of their neighbors, and finally completed
their work unmolested. Indeed, a perfect harmony
was established between the parties, which I never
saw interrupted by a single quarrel during the time
they remained my tenants."

This little bird is half-brother to our Purple Finch,
which inhabits Canada and the Northern and Eastern States during Summer, and the Middle and Southern States during Winter. The latter, however, does
not possess the mild and peaceable disposition of the
former, but is very quarrelsome and noisy even among
themselves. When feeding, as they often do, in
small flocks, upon the same trees, if one should happen to approach too near the spot where another is

cropping the tender buds, a difficulty mostly ensues, in which the weaker party is compelled to retire.

But the most remarkable and noteworthy member of the Finch family is the Crossbill. The singular form of the bill, and the peculiar manner in which it collects its food, give it a more than usually interesting character. There are two species, differing somewhat in their plumage, as well as in the locality in which they are found. The Common Crossbill, which appears to be the most abundant, inhabits during winter the pine forests of the Northern and Middle States, extending its migrations as far south as Maryland. They congregate in small flocks or families, and glean among the ripened cones of the Firs and Pines, where they find an abundant supply of nutritious food in the sweet kernels, which they detach from the husks with great dexterity. At first sight the bill of this bird appears like a deformity, but upon further observation we find that for the purpose to which it is applied by the owner, no better form could have been adopted ; and we are obliged to confess that Nature, in thus deviating from the usual form, understood well her own purposes, and that instead of its being a monstrosity, it is only another striking proof of the wisdom and skill of an Omnipotent Creator. Their food, although consisting principally of the seeds of the Pine and other cone-bearing trees, is not by any means confined to them. When in the vicinity of an orchard, if there is any fruit, they are sure to be among it, cutting the apples to pieces to get at the seeds, of which they

are very fond. They move about with great nimbleness among the close, thick-set branches of the Firs,
and when perched upon a cone will often stand upon
one foot while they use the other in conveying the
food to the mouth, somewhat after the manner of the
Parrots. The plumage of the males is mostly a fine
light yellowish red, intermingled with olive brown;
they vary much according to age and other circum-

White-winged Crossbill.

stances, and it is
very difficult to pro-
cure two birds in one
flock that are pre-
cisely similar. In
the White-winged
Crossbill this differ-
ence is not so ob-
servable, the mark-
ings being always
more distinct and
the colors stronger.
With the exception
of the wings and tail,
the whole body is of
a rich crimson, in-
terspersed with olive
and black; the wings and tail are black, the former
being crossed with bars of pure white. This bird
does not winter so far to the southward as the former
species, at least it seldom makes its appearance in the
latitude of Philadelphia, appearing to enjoy a colder
and more northern or mountainous range. Its habits

are similar to those of the former, its food and the manner of collecting it being the same.

We will now conclude our observations among the Finches by noticing three more birds, which, for brilliancy of coloring, are perhaps unsurpassed by any of our feathered friends, unless it be the Nonpareil.

The first is the Cardinal Grosbeak, that gay, active, and showy bird, which we sometimes see during a snow-storm, in company with the Snow Birds, flitting about among the trees and bushes, uttering its sharp chirp, and seeming to enjoy rather than lament the rigors of the season. The plumage of this bird, under whatever circumstances it is viewed, must ever render it an attractive object. Whether seen through the deep foliage of Summer, busily engaged with its domestic concerns, or whether in a more inclement season it rambles with freedom over the snow-clad fields and through the leafless woods, its imposing form, the lengthened crest by which its head is surmounted, and its livery of fiery red, cannot fail to arrest the eye. In richness of plumage and strength of song, it is probably not surpassed by any of the other American Grosbeaks. There are various names by which it is known in the different sections of country it inhabits, such as Red Bird, Virginia Nightingale, Cardinal Bird, etc. It is seldom seen to the eastward, or north of the southern boundary of New York. Southward from Maryland as far as Texas, it appears to be a constant resident; some individuals remaining during winter in the warmer parts of Penn-

8 *

sylvania and New Jersey. Audubon says of this bird : " Its song is at first loud and clear, resembling the finest sounds produced by the flageolet, and gradually descends into more marked and continued cadences until it dies away in the air around.   During the love season, the song is emitted with increased emphasis by this proud musician, who, as if aware of his powers, swells his throat, spreads his rosy tail, droops his wings, and leans alternately to the right and left, as if on the eve of expiring with delight at the delicious sounds of his own voice.   Again and again are those melodies repeated, the bird resting only at intervals to breathe.   They may be heard from long before the sun gilds the eastern horizon, to the period when the blazing orb pours down its noonday floods of heat and light, driving the birds to the coverts to seek repose for awhile.   Nature again invigorated, the musician recommences his song, when, as if he had never strained his throat before, he makes the whole neighborhood resound, nor ceases until the shades of evening close around him."

The Blue Grosbeak is also an inhabitant of the southern portion of the United States, but, unlike its brother the Cardinal, is a shy, modest species, retiring to the deep recesses of some secluded spot, where the footsteps of the white man are seldom seen.   Here, by the borders of some stagnant pool, where the poisoned fumes exhaled by decaying vegetation are filling the air, are the favorite haunts and the chosen summer dwelling of this beautiful bird. It is rarely seen north of Virginia, although individuals have been obtained in New Jersey and Pennsylvania; southward from this it is more abundant, extending as far as Texas. It has also been seen in considerable numbers among the Rocky Mountains. Its plumage much resembles that of the Indigo Bird, but the blue upon the head and throat is much finer and lighter. Although the nest of this bird is generally built near the ground, either in a low bush or a tuft of rank grass, it is observed that the male, which is possessed of a fine song, seldom or never utters more than a monotonous chirp when near it; but, retiring to the top of a tall detached tree, he will there indulge for some time in a succession of melodious strains.

We now present to your notice a bird which is pre-eminently beautiful, in every sense in which the term is applicable. This is the Scarlet Tanager. Look at him, with his gracefully formed body clothed in the most brilliant and glowing scarlet, and his wings and tail of jetty black, as he sits upon a tree with a strong light falling upon him, or as he gambols among the thick foliage, uttering his simple plaintive note, and we shall behold one of the most lovely and attractive objects which *our* feathered world can afford. Every one should be familiar with the habits as well as the appearance of this elegant bird. It is widely scattered over the United States during the summer months, and although seeming to have a decided preference for the woods, it may sometimes be seen about the farm-house and in the orchard, where he occasionally builds his nest. This

is a very slight structure, formed of the dry stalks
of flax or grass, and so loosely put together that the
light may easily be seen through it. The eggs are
mostly three, of a dull blue color, spotted with brown
or purple. Scarcely anything can exceed the attachment which these birds manifest for their young, as
the following incident related by Wilson will show :

" Passing through an orchard one morning, I
caught one of these young birds, that had but lately
left the nest. I carried it with me about half a mile,
to show it to my friend, Mr. William Bartram ; and,
having procured a cage, hung it up on one of the
large pine trees in the botanic garden, within a few
feet of the nest of an Orchard Oriole, which also
contained young ; hopeful that the charity or tenderness of the Orioles would induce them to supply the
cravings of the stranger. But charity with them, as
with too many of the human race, began and ended
at home. The poor orphan was altogether neglected,
notwithstanding its plaintive cries ; and, as it refused
to be fed by me, I was about to return it back to the
place where I had found it, when, toward the afternoon, a Scarlet Tanager, no doubt its own parent,
was seen fluttering round the cage, endeavoring to
get in. Finding this impracticable, he flew off, and
soon returned with food in his bill, and continued to
feed it till after sunset, taking up his lodgings on the
higher branches of the same tree. In the morning,
almost as soon as day broke, he was again seen most
actively engaged in the same affectionate manner ;
and, notwithstanding the insolence of the Orioles,

continued his benevolent offices the whole day, roosting at night as before. On the third or fourth day, he appeared extremely solicitous for the liberation of his charge, using every expression of distressful anxiety, and every call and invitation that Nature had put in his power, for him to come out. This was too much for the feelings of my venerable friend; he procured a ladder, and, mounting to the spot where the bird was suspended, opened the cage, took out the prisoner, and restored him to liberty and to his parent, who, with notes of great exultation, accompanied his flight to the woods!"

The Tanager family embraces not only the Tanager proper, but the American Wood Warblers. The latter are most numerous in the Northern continent, as the former abound in the Southern.

One of the most beautiful and interesting of the Wood Warblers is the American Redstart. This gay little bird is more noticeable in the early months of Spring, when it may be seen in company with many

American Redstart.

other species of its kind, nimbly flitting from tree to tree, and from twig to twig, in search of insects, of which it is a very expert hunter; it will sometimes pursue a swarm of small flies for a long distance, during which the snapping of its bill may be distinctly heard. In the deep shade of the woods the beauty of its markings shows to great advantage; the jetty black, which is the predominating color, contrasting finely with the streaks and bands of orange and vermilion on the sides, wings, and tail.

This bird and the little Blue-gray "Fly-catcher," differ slightly from the greater number of the Wood Warblers in their more fly-catching habits. All are diminutive birds, generally very abundant in the Middle States early in the Spring, but mostly retiring to the North during Summer to rear their young. Their principal appearance is in the morning just after sunrise, when every tree seems tenanted with them, all actively engaged in making a morning meal; this consists of insects and their larvæ, of which they devour great quantities. Many of them are expert fly-catchers, nimbly darting after the passing flies, while others are equally dexterous in clambering among the branches of the trees, hanging sometimes head downward, and holding on with one foot, and stretching their little necks in all directions in search of a favorite worm. Although these transient visitors are, with some exceptions, nearly destitute of song, yet among them are to be found some of our most beautifully plumaged birds.

The Yellow-poll Warbler, whose shrill notes are heard so constantly, during Spring and Summer, from almost every grove, and not unfrequently from the trees which surround the farm-house, and the

Maryland Yellow-throat, are perhaps the most familiar representatives of the family. The former is clad in a livery of brilliant golden yellow, spotted on the sides and breast with lengthened marks of chestnut orange. It builds a curious nest, suspended mostly among the forked branches of a low bush in the densest part of a thicket; it is composed of flax or tow, which is well woven into a neat little bag, and lined with hair or the soft down from various plants; the whole is well fastened to the stems from which it is hung, by the threads of tow or flax being tightly twisted about them. While the female is sitting, the male bird will often feign lameness, in order to draw away the attention of an intruder from the objects of his affectionate care.

The Maryland Yellow-throat is the humble and retired occupant of the low bushes and briers which are generally found growing on the banks of small streams and in wet marshy places: here it twitters out its sweet and animated song of "Whitit'iti! Whitit'iti!" repeating it in rapid succession for a few times, as it rambles among the branches where its food

Upper fig.—Yellow-poll Warbler.
Lower fig.—Magnolia Warbler.

is lurking. Its ambition seldom tempts it to leave the vicinity of the chosen spot where its nest is hung, nor to fly much above the level of the Alder and Hazel tops which surround its dwelling. It will, however, sometimes stray into the fields of growing grain, where it undoubtedly renders great service by the destruction of a multitude of noxious insects.

Both of these little birds are selected by the female Cow Bunting as foster-parents, to whom she commits the care of her young, by dropping her eggs in their nests. This singular and unnatural habit, of which we may say more in a future chapter, we believe does not exist in any other bird but the European Cuckoo, and is a curious instance of the wonderful variety to be seen everywhere in the works of an Omnipotent Deity.

There are above twenty other species of these lovely little birds, some of which are very conspicuous for their beauty; among them are the Blackburnian Warbler, Hooded Warbler, Magnolia Warbler, Cerulean Warbler, Cape May Warbler, and the Mourning Warbler. The latter is so named in consequence of the peculiar

Yellow-rumped Warbler.

markings of the head and neck, they being of a

beautiful leaden color, with bands of black upon the lower part of the throat. Most of these may be seen for a few weeks early in Spring, but it is difficult to distinguish between them, as they often frequent the tops of the tallest forest trees; at other times they have been known to enter the city, and hop about the shrubbery of the gardens. At the most they are only known to us as the transient occupants of our fresh-budding groves, the cooler atmosphere of the mountains to which they retire, being more congenial, and more favorable for the purposes of incubation.

Closely connected with the Wood Warblers, is the family of the *true* Warblers. As an illustration of these, let us take the famous Nightingale of Europe, whose powerful and melodious voice excites the wonder and praise of the listener. That such a long-continued succession of loud, clear, and musical notes can be produced by a bird of such small dimensions, is truly astonishing. It is no less remarkable for the great variety in the tones than for their peculiar clearness and melody. In order to illustrate this point, some writer has attempted to reduce the notes to plain English, — a copy of which we here place before our readers:

"Tioû, tioû, tioû, tioû, — Spe, tiou, squa, — Tiô, tiô, tiô, tiô, tiô, tio, tio, tix, — Coutio, coutio, coutio, coutio, — Squô, squô, squô, squô, — Tzu, tzu, tzu, tzu, tzu, tzu, tzu, tzu, tzu, tzi, — Corror, tiou, squa, pipiqui, — Tozozozozozozozozòzozozozo, zirrhading! Tsissisi, tsissisisisisisisis, — Dzorre, dzorre, dzorre, dzorre, hi,

— Tzatu, tzatu, tzatu, tzatu, tzatu, tzatu, tzatu, dzi,
— Dlo, dlo, dlo, dlo, dlo, dlo, dlo, dlo, dlo, — Quio,
tr-rrrrrrrr itz,—Lu, lu, lu, lu, li, li, li, li, liê, liê, liê,
liê, — Quio didl li lulylie,— Hagur, gurr, quipio!—
Coui, coui, coui, coui, qui, qui, qui, gui, gui, gui,
gui,—Goll, goll, goll, goll, guia hada doi,—Couigui,
horr, ha diadia dill si!—Hezezezezezczezezezezczezcze-
zezezeze cowar ho dze hoi, — Quia, quia, quia, quia,
quia, quia, quia, quia, ti, — Ki, ki, ki, ïo, ïo, ïo
ioioioio ki,—Lu ly li le lai la, leu lo, didl ïo quia,—
Kigaigaigaigaigaigaigaigai guiagaigaigai couior dzio
dzio pi."

CHAPTER V.

INSESSORES: *PASSERES, CLAMATORES, AND OSCINES.*

THE FLY-CATCHER — PEWEE — KING BIRD — GREAT CRESTED
FLY-CATCHER — WOOD PEWEE — WREN — GREAT CAROLINA
AND WINTER WREN — CHICK-A-DE-DE — BROWN CREEPER.

THE resemblance which exists between the Swallows and the Fly-catchers, both in their formation and some of their habits, will at once be recognized by comparison. But differences will also be noticed sufficient to mark them as belonging to entirely distinct families. The great powers of flight which appertain to both are differently employed. The former seeks its insect food upon the wing, in a long-continued ramble over hill and dale, meadow and lake, in which it seems to be more bent upon enjoying the pleasures of the chase, than upon merely gratifying its appetite; while the latter contents itself with perching upon a twig, a fence-stake, or a tall stalk, quietly awaiting the approach of some favorite insect, when, quick as thought, it sallies forth in pursuit, generally securing it in one wild sweep, and returning quickly to its former stand-point, to watch for the arrival of a fresh victim.

In North and South America the Fly-catchers are replaced by a family whose habits and manners are entirely similar, but whose structure places them in a widely different position in the system. Their singing organs being of the more imperfect type, they are assigned to the suborder Clamatores, while the true Fly-catchers, like the Swallows, belong to the Oscine suborder. These Tyrants, or Tyrant Fly-catchers, as they are called, are abundant in almost every section of the country; there are few persons who have not had the opportunity of being familiar with the notes and appearance of many of them.

Among the first birds which cheer our hearts at the approach of Spring, is the Pewee Fly-catcher, his soft, sweet, and not unmusical voice often sounding through the leafless grove long before the last traces of Winter have yielded to the softening sunbeams. The song of the Pewee is a sure and reliable prognostic of the coming of that lovely season when the earth again clothes herself in her beautiful garments, and the air resounds with Nature's sweetest music. The social and familiar habits of this plain and modest little bird, as well as his confiding trust in man, must ever secure for him a conspicuous place in our affections, and entitle him and his little property to our earnest and zealous protection. This familiarity, however, sometimes subjects him to being made the mark of cruel and unthinking boys, who, with that wilful propensity for throwing stones which seems to be part of a boy's nature, are so reckless of consequences as to tease and torment the poor little

Upper fig —Wood Pewee. *Lower fig.* —Tyrant Fly-catcher, or King Bird.

9 *

bird, until one, more "lucky" than the rest, strikes
the deadly blow. The writer still remembers with
what sorrowful feelings, when a boy, he once held in
his hand the body of a Pewee, which with a random
toss of a stone he had deprived of life. Could all
children feel as he then felt, how wrong it is wan-
tonly to destroy that life which all have an equal
right to enjoy, they would cease to make sport of it,
and this charming little songster would possess to the
full that security to which he is justly entitled.

The Pewee often returns to a favorite summer resort for several successive years, occupying the same
nest each season, merely repairing the injuries which
it has received during the Winter. Audubon speaks
of his having found the same pair of birds occupying
a familiar nook in an old cavern which he had been
accustomed to visit for a number of years. At one
time he fastened to the legs of each of a brood of
young birds, the offspring of this pair, a ring of silver thread; these they carried about with them for
some time, and in the following Spring two of them
were seen in the same vicinity, still wearing the silver ring.

The King Bird, or Tyrant Fly-catcher, is also a
familiar summer visitant. Although by no means a
large bird, he is nevertheless gifted with a degree
of courage that would do justice to the largest of our
feathered race; and being remarkably quick and active upon the wing, he becomes a formidable enemy
to such of his neighbors as have the temerity to en-
croach upon his dominions. In the early part of the

Summer his jealous and quarrelsome disposition is most apparent. While his mate is occupied with her domestic concerns, he is ever watchful for the appearance of intruders, and any attempt to be sociable is repelled with little ceremony. The Eagle, the Hawk, and the Crow, although greatly his superiors in size and strength, are equally the objects of his animosity, and no sooner does one of them make his appearance, than our hero sallies forth to give him battle; and mounting above him, he darts down upon his back with the swiftness of an arrow, and by repeated pecks with his sharp, powerful bill, from which his less active foe finds it difficult to escape, he soon remains master of the field, having driven the intruder quite out of the neighborhood. There is, however, one bird, which, although no larger nor stronger than himself, has often proved too much for him; this is the Purple Martin. His superior quickness upon the wing enables him to evade the sharp blows of the King Bird's bill, and very frequently to get the mastery of him and drive him off; sometimes a long and obstinate contest between them ends in the death of the latter.

Notwithstanding the fondness of the King Bird for bees and sometimes for fruit, he is among the best of the farmer's friends. No Hawk will venture near a barn-yard while he is about, while the swarms of noxious insects which he daily destroys, together with other little services for which we are indebted to him, strongly recommend him to our special care and protection.

The extent of country over which he roams is very wide, reaching from Texas to Canada, and as far west as the Columbia river. In Florida his place is supplied by the Piping Fly-catcher, which he so nearly resembles that they might by some be mistaken for the same bird, being possessed of the same active and courageous disposition when intruded upon by a stranger.

We have also abounding in our woods during the summer months the Great Crested Fly-catcher and the Wood Pewee, the former a noisy, active fellow, often frequenting the orchard about cherry time, the latter a sprightly little bird about the size of a Sparrow, whose sweet notes of "Powee! Powee! Petoway!" prolonged with a mournful accent, may be heard from morning till evening; even during the heated hours of noon, when most other birds are silent, this little songster still utters his plaintive ditty with a sweet earnestness that cannot fail to attract attention.

We will now take up the families of the more perfect singing birds, though with regret that our limits will not permit a foray into the lands of sun and flowers, the tropical home of the lovely Cotingas, which are represented by a few species in the southwestern regions of our country.

We will first notice the Wrens and Titmice. With the former almost every one has some acquaintance. There are several very beautiful species inhabiting the country west of the Mississippi, but our knowledge of them is but limited. Of those further east-

ward we shall take some notice.   Who does not love
the first sight of the House Wren, as he returns to
us after his long winter rambles in the south?   His
sweet and sprightly song is the very key-note of
Spring, speaking of cloudless skies and verdant fields,
of balmy air and music from the groves, of frolics·
among the wild flowers and rambles with the butter-
flies; it speaks of love and joy and happiness among
the myriad hosts of merry choristers, who are winging their way from tropical climes to join in the
grand harmony of Nature.   Let us read what Wilson
says of the Wrens:

"This well-known and familiar bird arrives in Pennsylvania about the middle of April, and about the

8th or 10th of May
begins to build its
nest, sometimes in
the wooden cornice
under the eaves, or
in a hollow cherry-
tree, but most com-
monly in small box-
es, fixed on the top
of a pole, in or near
the garden, to which
he is extremely par-

House Wren.

tial, for the great number of caterpillars and other
larvæ with which it constantly supplies him.   If all
these conveniences are wanting, he will even put up
with an old hat nailed on the weather-boards, with a
small hole for entrance; and, if even this be denied

him, he will find some hole, corner, or crevice about the house, barn, or stable, rather than abandon the dwellings of man. In the month of June, a mower hung up his coat under a shed, near a barn; two or three days elapsed before he had occasion to put it on again; thrusting his arm up the sleeve, he found it completely filled with some rubbish, as he expressed it, and, on extracting the whole mass, found it to be the nest of a Wren completely finished, and lined with a large quantity of feathers. In his retreat he was followed by the little forlorn proprietors, who scolded him with great vehemence for thus ruining the whole economy of their household affairs.

"This little bird has a strong antipathy to cats; for, having frequent occasion to glean among the currant-bushes, and other shrubbery in the garden, those lurking enemies of the feathered race often prove fatal to him. A box fitted up in the window of the room where I slept, was taken possession of by a pair of Wrens. Already the nest was built, and two eggs laid, when one day, the window being open, as well as the room door, the female Wren, venturing too far into the room to reconnoitre, was sprung upon by Grimalkin, who had planted herself there for the purpose, and, before relief could be given, was destroyed. Curious to see how the survivor would demean himself, I watched him carefully for several days. At first he sung with great vivacity for an hour or so, but, becoming uneasy, went off for half an hour; on his return, he chaunted again as before, went to the top of the house, stable, weeping willow, that she might hear him; but, seeing no appearance of her, he returned once more, visited the nest, ventured cautiously into the window, gazed about with suspicious looks, his voice sinking to a low, melancholy note, as he stretched his little neck about in every direction. Returning to the box, he seemed for some minutes at a loss what to do, and soon after went off, as I thought, altogether; for I saw him no more that day. Toward the afternoon of the second day he again made his appearance, accompanied with a new female, who seemed exceedingly timorous and shy, and who, after great hesitation, entered the box. At this moment the little widower or bridegroom seemed as if he would warble out his very life with ecstacy of joy. After remaining about half a minute in, they both flew off, but returned in a few minutes, and instantly began to carry out the eggs, feathers, and some of the sticks, supplying the place of the two latter with materials of the same sort, and ultimately succeeded in raising a brood of seven young, all of which escaped in safety.

"Its food is insects and caterpillars, and, while supplying the wants of its young, it destroys, on a moderate calculation, many hundreds a day, and greatly circumscribes the ravages of these vermin. It is a bold and insolent bird against those of the Titmouse and Woodpecker kind that venture to build within its jurisdiction; attacking them without hesitation, though twice its size, and generally forcing them to decamp. I have known him to drive a pair of Swallows from their newly formed nest, and take immediate possession of the premises, in which his female also laid her eggs, and reared her young. Even the Bluebird, who claims an equal and sort of hereditary right to the box in the garden, when attacked by this little impertinent, soon relinquishes the contest, the mild placidness of his disposition not being a match for the fiery impetuosity of his little antagonist. With those of his own species who settle and build near him, he has frequent squabbles; and when their respective females are sitting, each strains his whole powers of song to excel the other."

Great Carolina Wren.

The Great Carolina Wren and the Winter Wren are also, both of them, lovely and interesting birds. The former frequents the banks of streams, shaded by thickly overhanging foliage, where it may be distinguished by its clear, musical note, resembling the words Sweet William, Sweet William, uttered in rapid succession, with an occasional interlude of "Chirr-up, Chirr-up." It may also be found frequenting damp rocky caves, and among old piles of rotten timber, where it picks up the larvæ of many a hurtful insect. The Winter Wren visits us in Pennsylvania from the north, just as the House Wren has left us for its tropical home. It sometimes passes the entire winter in the Middle States, where it may be seen hopping about the wood-piles and the fallen and decayed trunks of trees, with its tail erect, busying itself in singing its musical ditty, and picking up the bugs that may be lurking in the crevices of the bark. It disappears again early in Spring, and passes to the northward in company with the Snow Birds.

Upper fig.—Crested Titmouse.
Lower fig.—Black-capped Titmouse.

The Titmouse, like its cousin the Wren, is an active, cunning little creature, ever on the go, hop, skip, and jump, from branch to branch, head down or head up, as is most convenient, incessantly prying into the private affairs of the insect world, often laying waste the prospects of a promising family with one stroke of its bill; and hunting up the vermin with such untiring industry as fairly to win for him a conspicuous place among the farmer's friends. There are two species

with which we are familiar; the Black-capped Titmouse, or Chick-a-de-de, and the Crested Titmouse.
They are both constant residents in the Middle States,
Summer and Winter; but it is during the severity of
Winter that we are most accustomed to their appearance. They then assemble in small troops with the
Snow Birds and the little Spotted Woodpecker, and
entering the orchard, or the trees around the house,
they soon make themselves known by their incessant
chatter, and great activity in chasing each other from
tree to tree. The notes of the former, when thus
engaged, are very rapid, and uttered with considerable energy, bearing some resemblance to the words
" See, see, sweet, sevait, chick, chick-a-de-de." The
latter has, in addition to his lively twitter, a loud
whistle, which may be heard for hours together, repeated at intervals as though calling a dog. These
little birds are apt to build their nests in the deserted hole of a Woodpecker; but frequently, when
none such are to be found, they will work with great
perseverance until they have made one for themselves, even picking their way into the trunk or
branches of some of our hardest wooded trees. Associated with them may often be seen the Brown
Creeper, a plain, modest, unassuming little fellow,
whose utmost ambition seems to be to fill its stomach
with the dainty little morsels which it picks out from
the crevices and holes in the trees with its long sharp
bill.

CHAPTER VI.

INSESSORES: *SYNDACTYLI.*

DESCRIPTION OF THE NIGHT HAWK — WHIP-POOR-WILL — CHUCK-WILLS-WIDOW — BARN AND CHIMNEY SWALLOWS — ANECDOTE BY AUDUBON—PURPLE MARTIN—EDIBLE SWALLOW'S NEST.

IT is extremely interesting, in the study of Birds, to notice the connection which exists between tribes as well as species. We have spoken of the Hawk Owl as possessing peculiarities of form and habits belonging to two distinct families; we will now notice other instances which are no less remarkable in this respect. In the Night Hawk, the Whip-poor-will, and the Chuck-wills-widow, we observe the soft downy plumage and the muffled wings of the Owl, as well as its nocturnal habits, combined in many prominent points with the general structure of the Swallow. The wide mouth, the small sharp bill, slightly hooked, the short legs and small feet, the long sharp wing and wide expanding tail. With the Owls ends the division Raptores, and with the Night Hawk, etc., commences the order of Insessores.

The habits of the three birds above-named are extremely interesting. With the Night Hawk we are most familiar, as it is quite abundant everywhere, from Maine to South Carolina, and westward to the

Rocky Mountains. The name of this bird is in singular disagreement with its most marked characteristics, it being generally seen upon the wing in broad day, often when the sun is shining brilliantly, and mostly retiring to rest soon after dusk. It may frequently be seen flying over the steeples and tall chimneys of our most densely populated cities, and sometimes builds its nest upon the house-top. Its food consists of large insects, which it procures upon the wing. When engaged in their pursuit, its motions are very graceful and interesting, and as it glides around in endless gyrations, flinging itself with the most careless ease upon the bosom of every gale, now rising, and now, like an arrow, dropping on its prey, at intervals uttering a shrill scream, then darting off in a wild zigzag course, snapping up every insect that comes within its reach, its actions may be followed by the eye with no small degree of pleasure.

In Louisiana it makes its appearance from the south early in the Spring; here it spends several weeks of the time occupied in its migrations, and is seen sailing over the cotton and sugar plantations, picking up here and there an unlucky beetle, or gambolling wildly over the prairies, lakes, and rivers from morning until evening.

There is probably no other bird, except the Swallow, which can rival the Night Hawk in the beauty and ease of its aërial motions, abounding as they do in feats of the most wonderful agility. Sometimes it will raise itself several hundred feet in the most careless manner, crying louder and louder as it as-

10 *   H

cends, then instantly it will glide obliquely downward with astonishing rapidity, until within a few feet of the ground, when, with the quickness of thought, it expands its wings and tail to the utmost, thus checking its downward course, and darting off with wonderful swiftness for a short space, mounts again almost perpendicularly. So great is the muscular power of its wing, that these evolutions are continued for hours almost without rest.

While the Night Hawk seems to be very generally distributed over the territory of the United States lying north of Louisiana, the Whip-poor-will and Chuck-wills-widow are confined to much narrower limits, — the former not extending its migrations much north of New York and the southern parts of Maine, and the latter seldom being seen north of Virginia.

By some the Whip-poor-will has been confounded with the Night Hawk, but the difference in their habits marks them as distinct species; the fact that the latter retires to its roosting-place just as the former is emerging from its seclusion, may have led some careless observers to conclude they were the same. The Whip-poor-will is strictly a nocturnal bird, never appearing abroad by daylight except when forced by circumstances; but no sooner has the sun disappeared behind the western hills, and the shades of evening have closed around the thicket which gives it cover by day, than it bestirs itself, and peeps out upon the dim landscape over which the pale moon is casting a feeble glare. It is then that its sweet and sprightly notes are heard echoing upon the still air, "Whip-poor-will! whip-poor-will!" repeated in rapid succession for some minutes together. Then with a few wild sweeps through the air upon its noise-less wing, in pursuit of its insect prey, it alights per-haps upon the fence or wood-pile, or even upon the roof of the house, and again utters its soft but clear cry with great animation. Those who have listened to the song of this bird, flowing like a liquid stream of melody, can alone judge of the soothing and quieting influence which it possesses.

Chuck-wills-widow.

The habits of the Chuck-wills-widow are very sim-ilar to those of the Whip-poor-will, and are equally interesting. In the pine forests of South Carolina it

is abundant, where its familiar and oft-repeated cry
of "Chuck-wills-widow!" is kept up during a great
part of the night. It is impossible to find language
to convey a just idea of the impression which the
notes of this bird produce upon the mind. Imagine
ourselves in the midst of a southern forest; tall pines,
interspersed with oaks and other forest trees, occupy
the ground for many miles around, covering it with
a broad canopy of shade, with here and there a wide
opening vista, through which the light may penetrate.
The sultry air is beginning to feel the cooling effects
of the falling dew, — the sun has long since sunk to
his rest, — the tree-tops wave gently in the twilight
gale,—the feathered songsters that have tenanted the
air during the long day have retired to their nests,—
the bee hums no more with her busy wing, and all
Nature is seemingly gathered into a sweet repose,
over which the quiet moon reigns with a serene ma-
jesty. This lull, however, is but temporary, an in-
terregnum between the dominion of day and the em-
pire of night; soon the screams of the wild-cat are
heard in the distance, as she sallies forth in quest of
her evening meal; the hooting of some monstrous
owl, that sails like a dim spectre overhead, salutes
the ear; frogs, lizards, and other reptiles are hopping,
skipping, and jumping about our feet; the whole air
becomes tenanted with a numerous insect life; and
a mingled chorus of hum, buzz, and chirp, everywhere prevails. We pause at one of the beautifully
expanded vistas, through which the full-orbed moon
gently darts her silvery beams, and gaze in silent

admiration upon the beauty of the scene; suddenly a swift-winged, noiseless phantom sails across our track, and alights upon a tree near by; it is then that we will listen to one of the most singular notes that is heard by night. Even the soft, full-toned, and richly varied song of the Mocking Bird, with which it is often blended, cannot drown the sweetly cadenced voice of this plain and unobtrusive bird, as he sits and " Chucks" and " Chuck-wills-widow" away, during the live-long night.

The unmeaning name of Goat-sucker has been applied to various members of this family of birds, the ignorant inhabitants of the countries where they are found supposing that they sucked the milk from their flocks, which is not only improbable, but altogether absurd. There are many species found in various parts of the world, some of them being quite large, and some not less noisy. Upon these last has been bestowed the appropriate name of Night Jars.

Of the myriads of winged visitors which annually flock to our shores from the south, there is perhaps no more interesting and familiar species than the Swallows. With what pleasant and happy recollections is their arrival associated! Spring, with all its attendant beauty, follows hard in the track of these little aërial voyagers; and the bright flowers whose half-expanded buds have lain almost concealed beneath the lingering snows, only await the gentle fanning of their wings to open into bloom.

Every farmer's child, and almost every school-boy in town or country, is at home among the Swallows;

they are associated with his earliest recollections; he may forget the dull pages that months of painful study have scarcely fixed upon his memory; but the appearance of the Barn Swallow, his easy, skimming, graceful flight, as he darts over the meadow, the lake, or the stream, his sprightly twittering note, and his nest under the barn roof, are things which he cannot forget.

The Barn and Chimney Swallows are by some ignorant persons thought to be the same bird; but a wide difference exists between them, both as to their appearance and habits. The plumage of the former is beautifully varied with a brilliant and glossy blue-black on the upper parts, and a rich fawn or drab color below; the tail being deeply forked, with the two

Barn Swallow.

outer feathers nearly double the length of the others; while the latter is wholly of a plain mouse or slate color, with the tail nearly even, and each feather ending in a sharp point.*

---

* The differences between the Chimney and Barn Swallows are greater and more important than our author himself appears to have been aware of. The Chimney Bird is a Swift, and belongs to a family of Syndactyli near the

The Chimney Swallows, when performing their migrations, often assemble to the number of several thousands, and take possession of the trunk of some venerable tree which has been hollowed out either by fire or by natural decay. Here they will continue to roost for many nights in succession before dispersing to the various parts of the country where they are accustomed to breed. Audubon thus describes a rendezvous of this kind which was tenanted by about 8000 or 9000 Swallows at one time:

"Immediately after my arrival at Louisville in the State of Kentucky, I became acquainted with the late hospitable and amiable Major William Croghan and his family. While talking one day about birds, he asked me if I had seen the trees in which the Swallows were supposed to spend the winter, but which they only entered, he said, for the purpose of roosting. Answering in the affirmative, I was informed that on my way back to town, there was a tree remarkable on account of the immense numbers that resorted to it, and the place in which it stood was described to me. I found it to be a sycamore, nearly destitute of branches, sixty or seventy feet high, between seven and eight feet in diameter at the base, and about five for the distance of forty feet up, where the stump of a broken hollowed branch, about two feet in diameter, made out from the main

---

Night Hawks. The true place of the Swallows is not in the present Chapter, but near the Tanagers, in Chapter IV. They belong to the singing division (Oscines) of the order Passeres.   E. D. C.

stem. This was the place at which the Swallows entered.   On closely examining the tree, I found it
hard, but hollow to near the roots.   It was now about
four o'clock in the afternoon, in the month of July.
Swallows were flying over Jeffersonville, Louisville,
and the woods around, but there were none near the
tree.   I proceeded home, and shortly after returned
on foot.   The sun was going down behind the Silver
Hills;   the evening was beautiful;   thousands of
Swallows were flying closely above me; and three or
four at a time were pitching into the hole, like bees
hurrying into their hive.   I remained, my head lean-
ing on the tree, listening to the roaring noise made
within by the birds as they settled and arranged
themselves, until it was quite dark, when I left the
place, although I was convinced that many more had
to enter.   I did not pretend to count them, for the
number was too great, and the birds rushed to the
entrance so thick as to baffle the attempt.

"Next morning I was early enough to reach the
place long before the least appearance of daylight,
and placed my head against the tree.   All was silent
within.   I remained in that posture probably twenty
minutes, when suddenly I thought the great tree was
giving way, and coming down upon me.   Instinctively I sprung from it; but when I looked up to it
again, what was my astonishment to see it standing
as firm as ever.   The Swallows were now pouring out
in a black, continuous stream.   I ran back to my post,
and listened in amazement to the noise within, which
I could compare to nothing else than the sound of a

large wheel revolving under a powerful stream. It was yet dusky, so that I could hardly see the hour on my watch; but I estimated the time which they took in getting out at more than thirty minutes. After their departure, no noise was heard within, and they dispersed in every direction with the quickness of thought." *

---

* The Swallows are undoubtedly sociable creatures, seeming disposed at least to be neighborly, and often, when unmolested, manifesting an inclination to live upon terms of intimacy with us which are sometimes inconveniently familiar. Scarcely a farm-house exists but whose chimneys are appropriated to the summer occupancy of one or more families of Swifts.

Some years ago, at a nobleman's house in Scotland, a pair of Swallows built their nest upon the top bar of a clothes-screen which was hung against the wall in the porter's lodge; the young were hatched and flew away. Upon the first appearance of the Swallows the following year, a male bird again entered the apartment and surveyed the premises. Having satisfied himself, he went off, but soon returned with a companion, which at first appeared very shy and timid, but in a short time acquired as much assurance as its mate. They both forthwith set about building a new nest on a small ledge which had been prepared for them as near as possible to the place where that on the clothes-screen had been built, and which had been destroyed; as, while it remained, the screen was of course useless to the family.

In this nest three broods were reared as before, notwithstanding the almost constant presence of the porter and his wife, who lived and slept in the room. In the Spring of the third year, the male again made his appearance with another mate, evidently much younger than her predecessor.

There are some species of Swallows which are remarkable for the beauty of their plumage, as well as for the gracefulness of their flight. The Violet Green Swallow and the White-bellied Swallow,—the former an inhabitant of the Rocky Mountains, and the latter quite an abundant species in the Eastern and Middle States,—are both entitled to a high rank among our gay-plumaged birds. We have also the Purple Martin, a very familiar and welcome bird in the Spring. When seen at a distance, it appears to be wholly black, but upon closer inspection it will be found to glisten all over with the most pleasing metallic hues, changing from blue to green, and from violet to golden purple, according to the position in which it is seen.

The Swallows possess undoubtedly greater powers of wing than any other birds. The space passed over in a few minutes by one of these little fairies is astonishing. Take for instance the Barn Swallow, and endeavor to follow him with the eye through all his curves and zigzag lines, as he darts about over some new-mown field; so rapid are his movements, that the keenest and quickest vision is often baffled in the attempt to retain its hold upon him,—and yet he flits on untiringly, mounting and falling, skimming and sailing, until the eye tires of his endless circuit.

The old nest on the ledge was examined, but the young partner possibly desiring a new home, the clothes-screen, which was hanging in the same position it had occupied the first year, was selected for the nest, and soon the process of hatching and rearing the first brood was in progress.

Wilson, upon whose accuracy of observation we can safely rely, considers one mile in a minute as a true estimate of the ordinary speed of this bird; and upon this he bases a calculation to show over what extent of ground in a straight line our little friend would glide during his short life, allowing ten hours of each day as the time occupied by the bird in performing his evolutions. According to this estimate, he will, during the ten years of his existence, have passed over the incredible space of 2,190,000 miles, or 87 times the circumference of the globe.

The form of the nest built by the different species of Swallow varies much. We are all familiar with the frail tenement of sticks in which the Chimney Bird deposits her snow-white eggs, and the neat and comfortable nest of the Barn Swallow, which it perches upon a projecting rafter near the peak of the barn. But the nest of the Cliff Swallow is of remarkable construction, being shaped like a gourd with a neck, and is composed of little pellets of mud, deposited by the bird one after another, until the required shape and size are attained. These nests are generally attached to the sides of a rock or projecting cliff, or to the walls of a building, sometimes as many as hundreds together. Their thus congregating and living in flocks or families has given them in some localities the name of Republican Swallows.

In the islands of Java and Ceylon, and many others adjacent, is found a species called the Edible

Swallow, from the fact that their nests form an article of food very highly prized by the Chinese epicures.   These nests are regarded as a great delicacy,
and are so much in esteem that the finest of them,
it is said, will bring their weight in silver.   They
form a very important article of trade, as about
thirty thousand tons of Chinese shipping are employed in it.   The income arising from this singular
traffic is appropriated by the government as one of
its revenues.

The following interesting account of the habits of
these birds, and the method of obtaining the nests,
is from Stanley's "Familiar History of Birds."
"The two bird-mountains [in the island of Java]
are insulated rocks, hollow within and pierced with
a great number of openings.   Many of these openings are so wide, that a person can enter them with
ease; others are attended with more difficulty, and
some are too small to admit of intrusion; in these,
therefore, the poor little birds are alone safe from
robbery.   To the walls of these caverns the birds
affix their small nests in regular rows, and so close
that for the most part they adhere together.   They
construct them at different heights, from fifty to sixty
feet, sometimes higher, sometimes lower, according
as they find room; and no hole or convenient place,
if dry and clean, is left unoccupied; but if the walls
be in the least wet or moist, they immediately desert
them.   At daybreak these birds fly abroad from their
holes, with a loud fluttering noise, and in the dry

season rise so high into the atmosphere in a moment, as they have to seek their food in distant parts, that they are soon out of sight. In the rainy season, on the other hand, they never remove to a great distance from their breeding-places.

"About four in the afternoon they again return, and confine themselves so closely to their holes, that none of them are seen any more flying, either out or in, but those which are hatching. They feed on all sorts of insects which hover over stagnant waters, and these they easily catch, as they can extend their bills to a great width. They prepare their nests from the strongest remains of the food which they use, and not of the scum of the sea, or of sea plants, as some persons have supposed. They employ two months in preparing their nests; they then lay their eggs, on which they sit for fifteen or sixteen days. As soon as the young are fledged, people begin to collect their nests, which is done regularly every four months; and this forms the harvest of the proprietors of these rocks.

"The business of taking them down from the rocky ledges on which they are placed, is performed by men who have been accustomed from their youth to climb among these dangerous places. They construct ladders of reeds and bamboos, by which they are enabled to ascend to the holes; but if the caverns are too deep they employ ship-ropes. When they have got to the bottom of a cavern, they place bamboos, with notches in them, against the wall, if

11 *

these be sufficiently long to reach the nests, but if
not they stand on the ladders, and pull the nests
down with poles of bamboo made for that purpose.
This employment, which is very dangerous, sacrifices
the lives of many men, and particularly of thieves,
while attempting to rob the caverns at improper
seasons."

CHAPTER VII.

INSESSORES: *SYNDACTYLI AND ZYGODACTYLI.*

BELTED KINGFISHER—CALIFORNIA AND RED-HEADED WOOD-
PECKERS—A NARRATIVE OF THE CALIFORNIA WOODPECKER,
BY "KELLEY" — IVORY-BILLED, GOLDEN-WING, YELLOW-
BELLIED, AND DOWNY WOODPECKERS—CUCKOO—PARROTS
—ANECDOTE OF A PARROT, FROM GOSSE'S "NATURAL HIS-
TORY OF BIRDS."

By the banks of some quiet, running stream, or
smooth and glassy mill-pond, where the Willow, Hazel and other shrubs dip their branches into the
sleeping waters, may often be heard a shrill, chatter-
ing note, much resembling the sound of the watch-
man's rattle, which falls with pleasing effect upon
the ear, as it gently dies away in the distance. This
is the note of the Belted Kingfisher, which our presence has just started from his perch near by. He
flies some distance up or down the stream, where he
selects a fresh stand-point, from which he intently
eyes the motions of the finny tribes below, until one
suited to his taste comes within the range of his
deadly aim, when with a sudden winding sweep he
darts below the surface, and seizing it with his powerful bill, bears it away to his perch, and immediately
swallows it whole.

This singular and not inelegant bird is a lone representative of its tribe in the United States; but being abundant wherever fresh water and good fishing are to be found, it has become quite familiar, occupying as prominent a place in our Natural History, as the pretty little European species does in the rural landscapes of Great Britain. The form and

Belted Kingfisher.

appearance of the Kingfisher are peculiar. A long, sharp, and powerful bill; a large head, surmounted by a crest that adds fierceness to its look; a thick neck and robust body, but rather small in proportion; wings ample; legs very short, and feet small. The upper parts of the plumage are bluish lead color, lower parts mostly white; in the male a band of black crosses the upper part of the breast; in the female the blue tint is not so perceptible, and the band across the breast is reddish brown, the belly being girted with a broad belt of the same color.

Its favorite places of resort are near inland streams,

Carolina Parrots or Parrakeets.

lakes, and mill-ponds, especially where a clayey or gravelly bank rises to some height above the water's edge; here the male and female assist each other in digging out a hole, running horizontally to the depth of four or five feet, and about one or two feet below the surface of the ground. This hole, which is just large enough to admit the body of the bird, is widened toward the extremity into an oven-shaped apartment, of sufficient size to allow of the birds turning freely about; here the nest, which is composed of a few sticks and feathers, is placed. The female mostly lays six pure white eggs, which she hatches in about sixteen days, the male taking his turn with his mate in the process of incubation. To this hole the same pair will sometimes resort for many successive years.

We will now endeavor briefly to describe some of the most prominent and familiar members of the interesting, numerous, and widely spread family of the Woodpeckers. With them commences the fourth order, Scansores or Zygodactyli, the Climbers. If we examine closely, we will find that the peculiarities of conformation of this order are very marked, and display in a wonderful degree the wisdom of the Creator in supplying His creatures with means precisely adapted to their wants. The food of the Woodpecker consists principally of insects and their eggs, which are deposited beneath the bark of decayed trees. In order to obtain these, it is gifted with a large, heavy looking, hammer-shaped head, and rather a long, sharp-pointed, and powerful bill, with which it strips away the bark by repeated blows, until it has uncov-

ered the object of its search. Sometimes the insects
have hollowed out for themselves a cavity beneath
the bark, extending for some distance into the wood
of the tree. These it dislodges by means of its
long tongue, which is barbed at the extremity, and
capable of being protruded to a great distance beyond
the point of the beak. The tongue is supported by
a series of small bones and cartilages, which find
their origin upon the forehead on each side of the
base of the bill. At first they lie pretty close together, but soon separate gradually, each passing
round the back part of the head, and entering the
mouth immediately below the ear, come together at
a point near the base of the bill. That part of the
tongue which lies between this point and the end of
the bill, is of a fleshy, worm-like appearance, and
ends in a slender, bony point, armed on either side
with sharp prickles, directed backward, but not capable of being moved forward. This barbed point is
particularly serviceable in drawing out from their
close concealment the heavy larvæ, which sometimes
measure two or three inches in length.

The protrusion of the tongue is produced by the
action of a pair of muscles, secured to the lower jaw
near the base of the lower mandible, and running
backward nearly the entire length of the bony process of the tongue. The position of the different
parts, and the singular structure of this important
member, will be better understood by reference to
the figure on the next page. With the bill it also
digs out of the solid wood a hole in which to raise

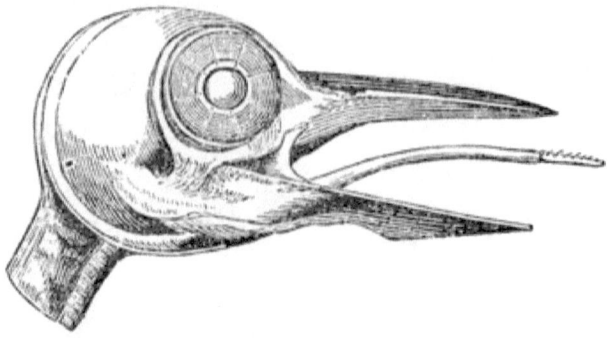

its young; these holes vary in depth, according to circumstances, from six inches to nearly three feet. In its usually upright position against the body of the tree, it must necessarily lie pretty close in order to work to advantage; hence the legs are short and muscular, and the toes, which are arranged two before and two behind, are well calculated to take a firm hold upon the bark and retain it for a long time. This upright position of the body is also more steadily maintained by the assistance of the tail, which is somewhat wedge-shaped, the more central feathers being more rigid, and having the shafts very thick, and stout, and sharp pointed, which, when placed against the trunk of the tree, serve as a support. The flight of the Woodpeckers is also singular, consisting of a frequent repetition of a few rapid and powerful strokes of the wings, which produce an undulating motion through the air, the body rising and falling with the alternate opening and closing of the wings. They are sprightly and active in their motions, alighting upon a tree and moving around the whole circumference, backward and forward, or ascending spirally

12

by a series of short leaps, occasionally stopping to give a few raps, to discover the lurking-places of insects beneath the bark; then on he goes, here and there picking up a dainty morsel, until being satisfied that he has done with the tree, he passes to another.

The United States are particularly favored in possessing a large number of species of this beautiful tribe. The California Woodpecker is justly celebrated for the richness of its plumage, as well as for some of its singular habits. We do not know of any other species that is so provident as to lay up stores for the winter. This propensity has been attributed to several others, but in the bird before us we have positive evidence to that effect from those who have witnessed its operations. The following interesting narrative, taken from Kelley's " Excursion to California," will throw some light upon the subject:

Woodpeckers.

" In stripping off the bark of this tree, I observed it to be perforated with holes, larger than those which

a musket ball would make, shaped with the most accurate precision, as if bored under the guidance of a
rule and compass, and many of them filled most
neatly with acorns. Earlier in the season I had re-
marked such holes in most of all the soft timber, but
imagining that they were caused by wood insects, I
did not stop to examine or inquire; but now finding
them studded with acorns, firmly fixed in, which I
knew could not have been driven there by the wind,
I sought for an explanation, which was practically
given me by Captain S——'s pointing out a flock of
Woodpeckers, busily and noisily employed in the
provident task of securing the winter's provision.
For it appears that this sagacious bird is not all the
time thriftlessly engaged in 'tapping the hollow
beech tree' for the mere idle purpose of empty sound,
but spends its summer season in picking these holes,
in which it lays its store of food for the winter, where
the elements can neither affect nor place it beyond
their reach; and it is regarded as a sure omen that
the snowy period is approaching, when these birds
commence stowing away their acorns, which otherwise might be covered by its fall. I have frequently
paused from my chopping, to watch them in the
neighborhood, with the acorns in their bills, half
clawing, half flying around the tree, and have admired the adroitness with which they tried it at different holes until they found one of its exact calibre,
when, inserting the pointed end, they tapped it home
most artistically with the beak, and flew down for
another.

" But the natural instinct of this bird is even more remarkable in the choice of the nuts, which are invariably found to be sound, whereas it is an utter impossibility, in selecting them for roasting, to pick up a batch that will not have a large portion of them unfit for use, the most smooth and polished frequently containing a large grub generated within. Even the wily Digger Indian, with all his craft and experience, is unable to arrive at anything like an unerring selection, while in a large bagful that we took from the bark of our log, there was not one containing the slightest germ of decay."

This Woodpecker appears to be very abundant, occupying a corresponding position with the well-known Red-headed species so common to the eastward. They also somewhat resemble each other in their plumage, the preponderating colors in both being black, white, and crimson.

The Ivory-billed Woodpecker is the largest species found within our territory, measuring twenty-one inches in length. It is an inhabitant of the Southern and Western States, and notwithstanding its somewhat awkward look, is certainly a noble and majestic bird. Spurning the low occupation of seeking his food among stunted trees and bushes, or upon prostrate logs and fence-rails, he leaves this humble game to the smaller fry of his tribe, and spreads his ample wings among the tall cypress and pines which cover vast areas of swampy ground in the Southern States. Here, amidst the security of these almost inaccessible forests, he regales himself upon the myriads of

insects which ever infest those noble trees. Wilson
says that, " Wherever he frequents, he leaves numerous monuments of his industry behind him. We
there see enormous pine trees, with cart-loads of bark
lying around their roots, and chips of the trunk itself, in such quantities as to suggest the idea that
half a dozen axe-men had been at work there for the
whole morning."

The same author relates the following amusing account of one of these birds which he had captured
in a wounded condition, and carried with him for
some distance. " This bird was only wounded slightly
in the wing, and on being caught, uttered a loudly
reiterated and most piteous note, exactly resembling
the violent crying of a young child, which terrified
my horse so, as nearly to have cost me my life. It
was distressing to hear it. I carried it with me in
the chair, under cover, to Wilmington, N. C. In
passing through the streets, its affecting cries surprised every one within hearing, particularly the
females, who hurried to the doors and windows with
looks of alarm and anxiety. I drove on, and on arriving at the piazza of the hotel where I intended to
put up, the landlord came forward, and a number of
other persons who happened to be there, all equally
alarmed at what they heard. This was greatly increased by my asking whether he could furnish me
with accommodations for myself and my baby. The
man looked blank and foolish, while the others stared
with still greater astonishment. After diverting
myself for a minute or two at their expense, I drew

12 *

my Woodpecker from under the cover, and a general laugh took place."

The head of this bird is ornamented with a crest of long flowing plumes, which, upon the forehead, are jetty black, while those of the hinder part are a brilliant crimson; the remainder of the plumage is mostly black, with slight reflections of blue. A white stripe, commencing at the ear, runs down each side of the neck, and half way down the back. The secondary quills in the wings, as well as a part of the primaries, are also white.

Some of the most familiar species to the north and eastward are the Red-headed Woodpecker, Golden-winged Woodpecker, or Flicker, Yellow-bellied Woodpecker, and the Downy Woodpecker, or Sap-sucker. Of these, the Red-headed Woodpecker may be considered as the most richly colored, displaying in its plumage one of the finest contrasts that could well be formed. The whole head and upper part of the neck are of a deep crimson, set off below by pure white, and above by a glossy steel blue. The secondary quills in the wings, and a broad band across the rump, are also white. The beauty of this bird renders him an attractive mark for the sportsman, for which reason the species appears to be on the decrease, and we fear that the day is not far distant when it will be numbered among our scarce birds.

Although the Red-heads undoubtedly do great service to the farmer in ridding his orchard and forest trees of a great number of insects, yet we cannot conceal the fact that their indulging in a fond-

ness for fruit and green corn has given them a reputation anything but enviable.   The finest and ripest
of the fruit are generally selected to gratify their
desires; and so keen is their relish for the early pro-
ductions of the orchard, that a well-loaded cherry-
tree will sometimes be entirely stripped of its cher-
ries before the owner has fairly tasted them.   The
pear and the apple-tree are equally the objects of
their regard; and should one be molested during his
depredations upon these, he will coolly thrust his bill
into as fine an apple or pear as he can, and bear it
away in his flight to the woods.   Much of the mischief which is done to the young corn, which is attributed to the Blackbirds, is undoubtedly the work
of this Woodpecker, as he will strip off the husk
from the ear almost in a twinkling, and regale himself
at leisure with its juicy contents.   These depredations are, however, more the exception than the rule;
his natural food is insects, and the amount of these
which he annually destroys, will more than compensate for the fruit and corn with which he varies his
diet.   We would, therefore, recommend him to the
protection of every one.   He is a bright, sprightly,
and attractive companion during our country strolls,
and cannot fail to afford us pleasure wherever we
meet him.

The Golden-winged Woodpecker, or Flicker, as
he is commonly called, although not so conspicuous
for his beauty as the preceding, is nevertheless a
handsome and showy bird.   The upper parts of the
plumage are dull bluish-grey upon the head, shading

into drab on the back, where it is crossed by bars of black, caused by each feather having a crescent-

shaped mark of that color near its extremity; the rump and upper tail coverts are nearly white, a band of bright vermilion runs from ear to ear around the hinder part of the head. The throat and upper part of the neck are reddish - fawn, extending to a broad band of black which crosses the breast; below this it' is dull fawn, shading gra-

Golden-winged Woodpecker.

dually into white on the under tail coverts, and variously marked with spots and bars of black. But the chief beauty of the bird consists in the color of the under surface of the wings and tail, which is a rich golden-yellow. In consequence of this being mostly concealed, his general appearance is rather plain and homely.

By some, the Woodpeckers have been regarded as dull, sleepy birds, possessed of but little animation or activity; but let such go to the woods and watch the motions of the Flicker as he gambols through the leafy bowers; see how he revels in the delights of Spring, ever on the go, uttering at frequent intervals his loud, clear, and not unpleasant cry. See with what assiduous devotion he and his mate assist each other in picking a hole into the solid heart of some sturdy oak; listen to the strokes of their bills; see the chips how they fly, and then call them sleepy birds. And when the cares of a brood are devolving upon them, they ply their busy bills with renewed activity, searching every nook and cranny that comes in their way for the daintiest worms, which they bear away to their young. See one of these birds when pursued by a Hawk; just as he is almost within the talons of his rapacious foe, he suddenly dives into a hole near by, or in the absence of this, he alights upon a tree and plays bo-peep with his enemy around its trunk. It is truly laughable to see how he dodges his pursuer, and you would wonder at the Hawk for wasting his time over such nimble game.

Westward of the Rocky Mountains there is a Woodpecker found almost precisely similar to the above, except that the under surface of the wings and tail are orange-red, the shafts of each feather being bright vermilion.

We must here reluctantly close our observations on this interesting group, leaving it to our readers to pursue the study, as inclination leads them, among the wild woods, where they will find some of the species abundant at all seasons of the year.

The Cuckoo, although not strictly a climbing bird, belongs to the same order as the Woodpeckers, the arrangement of the toes and other characteristics

assigning to it that position. We have several species in the United States, the most abundant being
the Yellow-billed Cuckoo. This graceful and familiar
bird, being of somewhat a quiet and retiring disposition, frequents the most secluded and thickest part
of the woods, where its low and simple notes of
" Cowe, cowe, coo, coo, coo," may be heard, uttered
at first slowly, but gradually increasing in rapidity
until the syllables run together. When it becomes
more clamorous than usual. it is said to be a sign of
approaching rain, which in some places has conferred

Cuckoo.

upon it the title of Rain Crow. The Cuckoos of
America, unlike their European relative, invariably
build their own nests and rear their own young, and
do not seem to be lacking in the least degree in a
strong affection for their progeny. The species now
before us has been accused of sucking the eggs of

other birds, which we regret to say does not appear
to be an unjust charge; in other respects he certainly
bears a good character as a quiet and harmless bird,
rendering good service to the farmer by the daily
destruction of a great amount of noxious vermin.
He often visits the orchard and garden, where he
sometimes builds his nest. Being strictly a summer
bird, he leaves the Northern and Middle States early
in the Autumn, for a warmer climate, many passing
the Winter in Florida.

The next and last division of the Scansores which
we have to notice is the Parrots, well known for their
peculiar form, their singular habits, and the brilliant
coloring of their plumage. Although many species
of this group are found in various parts of the globe,
yet the Equatorial Regions must be considered as the
favorite resort for by far the greatest number. Here,
among the wild and majestic forests of towering palms,
or in the deep and tangled thickets of mimosa, where
the face of Nature is clothed in perennial verdure,
these gay-feathered birds make the air resound with
their loud discordant cries. Each country seems to
be possessed of varieties or subdivisions of the group
somewhat peculiar to itself. Thus, from the interior
of South America we have the splendid Macaws,
which are generally large birds, over three feet in
length, of which the tail makes up twenty-four inches,
and decked in the most glaring hues of scarlet, green,
blue, and yellow. From India and the adjacent islands come the superb Lories, arrayed in their coats
of fiery red; while from Australia we welcome the

snow-white or roseate plumage of the Cockatoos.   It
is almost impossible for us to form any adequate conception of the extreme gracefulness and beauty of
these birds when enjoying the freedom of their native forests; and although their colors may be considered by many as too gaudy, and presenting too
many abrupt and striking contrasts to yield to the
eye that degree of pleasure which a softer blending
is apt to convey, yet we think that few can gaze upon
the multiplicity of their forms, and the richness and
diversity of their gorgeous tints, without regarding
them as one of the most wonderful and beautiful
families of the whole feathered race.

One of the most singular faculties of the Parrots,
— which, however, does not belong to the whole
tribe, — is that of imitating the human voice, and
learning by rote, words and sentences, which they
will sometimes repeat upon very appropriate occasions, giving the impression that they are really
aware of their meaning.   This power is possessed
principally by the short, even-tailed, and less gorgeously colored species.

The following interesting account of a remarkable
bird, probably the Grey African Parrot, which possesses
the greatest imitative powers, is from Gosse's "Natural History of Birds."   It is an extract from a letter
to a gentleman from the sister of its owner:

"As you wished me to write down whatever I could
recollect about my sister's wonderful Parrot, I proceed to do so, only premising that I will tell you nothing but what I can vouch for having myself heard.

Her laugh is quite extraordinary, and it is impossible to help joining in it oneself, more especially when in the midst of it she cries out, 'Don't make me laugh so. I shall die, I shall die;' and then continues laughing more violently than before. Her crying and sobbing are curious; and if you say, 'Poor Poll! what is the matter?' she says, 'So bad! so bad! got a bad cold!' and after crying for some time will gradually cease, and making a noise like drawing a long breath, say, 'Better now!' and begin to laugh.

"The first time I ever heard her speak, was one day when I was talking to the maid at the bottom of the stairs, and heard what I then considered to be a child call out, 'Payne! (the maid's name) I am not well! I'm not well!' and on my saying, 'What is the matter with that child?' she replied, 'It is only the Parrot; she always does so when I leave her alone, to make me come back;' and so it proved; for on her going into the room the Parrot stopped, and then began laughing, quite in a jeering way.

"It is singular enough, that whenever she is affronted in any way, she begins to cry, and when pleased, to laugh. If any one happens to cough or sneeze, she says, 'What a bad cold!' One day when the children were playing with her, the maid came into the room, and on repeating to her several things which the Parrot had said, Poll looked up, and said quite plainly, 'No I didn't.' Sometimes, when she is inclined to be mischievous, the maid threatens to beat her, and she says, 'No you won't.' She calls the

cat very plainly, saying, 'Puss! puss!' and then
answers, '*Mew;*' but the most amusing part is, that
whenever I want to make her call it, and to that purpose say, 'Puss! Puss!' myself, she always answers
'*Mew,*' till I begin mewing, and then she begins calling puss as quick as possible. She imitates every
kind of noise, and barks so naturally, that I have
known her to set all the dogs on the parade at Hamp-
ton Court barking; and the consternation I have seen
her cause in a party of cocks and hens, by her crow-
ing and clucking, has been the most ludicrous thing
possible. She sings just like a child, and I have
more than once thought it was a human being; and
it was ridiculous to hear her make what one should
call a false note, and then say, 'Oh, la!'' and burst
out laughing at herself, beginning again in quite another key. She is very fond of singing, 'Buy a
Broom,' which she says quite plainly; but in the
same spirit as in calling the cat, if we say, with a
view to make her repeat it, 'Buy a Broom,' she always says, 'Buy a *Brush,*' and then laughs, as a child
might do when mischievous. She often performs a
kind of exercise which I do not know how to describe, except by saying it is like the lance exercise.
She puts her claw behind her, first on one side and
then on the other, then in front, and round over her
head, and whilst doing so, keeps saying, 'Come on!
Come on!' and, when finished, says, 'Bravo! beautiful!' and draws herself up. Before I was as well
acquainted with her as I am now, she would stare in
my face for some time, and then·say, 'How d'ye do,

ma'am?' this she invariably does to strangers. One day I went into the room where she was, and said, to try her, 'Poll, where is Payne gone?' and, to my astonishment, and almost dismay, she said, 'Down stairs.' I cannot at this moment recollect anything more that I can vouch for myself, and I do not choose to trust to what I am told; but, from what I have myself seen and heard, she has almost made me a believer in transmigration."

The only member of this large family found in the United States is the Carolina Parrot, or Parrakeet; which, although not so brilliantly attired as some of the species, is nevertheless a very beautiful bird, the predominating color of the plumage being a light green, tinged with purple on the wings. The head and upper part of the neck are rich yellow, with a patch of orange-red upon the forehead. Many years ago, before the Southern and Western States became thickly settled, this Parrot was very abundant in those parts, but we believe that it is now seldom found much to the eastward of the Mississippi river. It is represented as an active, sprightly bird, and very graceful in its motions upon the wing. In the Autumn, when the Cockle Bur (a very noxious weed) has ripened its seed, they assemble in vast flocks, and, resorting to the fields where it grows, they alight upon the plants, and plucking the burs from the stem with their bills, they take them in one claw, while with the bill they open it and take out the fruit. In this way, a single flock will, in a few days, entirely rid a large field of the ripened seed;

the root of the plant, however, being perennial, they do not exterminate it.

Audubon says they do not confine themselves to the Cockle Bur exclusively, but attack all kinds of fruit indiscriminately, on which account they are always unwelcome visitors to the planter. They are particularly destructive to the grain-stacks, upon which they alight in numbers sufficient almost to cover it, pulling out the straws and scattering it about, thus wasting as much as they eat. While thus occupied, the farmer has a good opportunity of taking vengeance upon them for their unwarrantable intrusion. When once fired upon, all the survivors will rise, shriek, fly around a few minutes, and then alight again upon the same spot. The gun being kept vigorously at work, almost the entire flock is sometimes destroyed. At each discharge, the living birds fly over their slain or wounded companions, shrieking as loudly as ever, but still returning to the stack to receive their measure of what the farmer would call retributive justice.

These birds roost in companies, occupying the large cavities which are found in the sycamore trees, clinging to the sides of the hole as close together as they can crowd, hanging on with their bill and claws. They can scarcely be said to have any nests, their eggs being laid upon a few pieces of rotten wood at the bottom of the holes in which they roost.

Alexander Wilson, that accurate and beautiful ornithological writer, gives such an interesting account of one of these birds, which he kept for some time in confinement, throwing so much light upon their peculiar manners, that we cannot forbear inserting it:

"Anxious to try the effect of education on one of those which I procured at Big Bone Lick, and which was but slightly wounded in the wing, I fixed up a place for it in the stern of my boat, and presented it with some Cockle Burs, which it freely fed on in less than an hour after being on board. The intermediate time between eating and sleeping was occupied in gnawing the sticks that formed its place of confinement, in order to make a practicable breach, which it repeatedly effected. When I abandoned the river, and travelled by land, I wrapped it up closely in a silk handkerchief, tying it tightly around, and carried it in my pocket. When I stopped for refreshment, I unbound my prisoner, and gave it its allowance, which it generally despatched with great dexterity, unhusking the seeds from the bur in a twinkling; in doing which it always employed its left foot to hold the bur, as did several others that I kept for some time. In recommitting it to 'durance vile,' we generally had a quarrel, during which it frequently paid me in kind for the wound I had inflicted, and for depriving it of liberty, by cutting and almost disabling several of my fingers with its sharp and powerful bill. The path through the wilderness between Nashville and Natchez is in some places bad beyond description. There are dangerous creeks to swim, miles of morass to struggle through, rendered almost as gloomy as night by a prodigious

13 *

growth of timber, and an underwood of canes and other evergreens; while the descent into these sluggish streams is often ten or fifteen feet perpendicular, into a bed of deep clay. In some of the worst of these places, where I had, as it were, to fight my way through, the Paraquet frequently escaped from my pocket, obliging me to dismount and pursue it through the worst of the morass before I could regain it. On these occasions I was several times tempted to abandon it; but I persisted in bringing it along. When at night I encamped in the woods, I placed it on the baggage beside me, where it usually sat with great composure, dozing and gazing at the fire till morning. In this manner I carried it upwards of a thousand miles in my pocket, where it was exposed all day to the jolting of the horse, but regularly liberated at meal times, and in the evening, at which it always expressed great satisfaction. In passing through the Chickasaw and Choctaw nations, the Indians, wherever I stopped to feed, collected around me, men, women, and children, laughing and seeming wonderfully amused with the novelty of my companion. The Chickasaws called it in their language 'kilinky;' but when they heard me call it Poll, they soon repeated the name; and wherever I chanced to stop among these people, we soon became familiar with each other through the medium of Poll. On arriving at Mr. Dunbar's, below Natchez, I procured a cage and placed it under the piazza, where, by its call, it soon attracted the passing flocks; such is the attachment they have for each other. Numerous parties frequently alighted on the trees immediately above, keeping up a constant conversation with the prisoner. One of these I wounded slightly in the wing, and the pleasure Poll expressed on meeting with this new companion was really amusing. She crept close up to it as it hung on the side of the cage; chattered to it in a low tone of voice, as if sympathizing in its misfortune; scratched about its head and neck with her bill; and both at night nestled as close as possible to each other; sometimes Poll's head being thrust among the plumage of the other. On the death of this companion, she appeared restless and inconsolable for several days. On reaching New Orleans, I placed a looking-glass beside the place where she usually sat, and the instant she perceived her image, all her former fondness seemed to return, so that she could scarcely absent herself from it a moment. It was evident that she was completely deceived. Always when evening drew on, and often during the day, she laid her head close to that of the image in the glass, and began to doze with great composure and satisfaction. In this short space she had learned to know her name; to answer and come when called on; to climb up my clothes, sit on my shoulder, and eat from my mouth. I took her with me to sea, determined to persevere in her education; but, destined to another fate, poor Poll, having one morning about daybreak wrought her way through the cage, while I was asleep, instantly flew overboard, and perished in the Gulf of Mexico."

CHAPTER VIII.

INSESSORES: *SYNDACTYLI.*

HUMMING BIRDS.

The number of species of Humming Birds known to Linnæus, and other early naturalists, was comparatively few; while, more recently, Lesson, who has been considered a great exponent of the family, has, in his works upon that subject, only figured and described about one hundred and ten. But through the means of various travellers who have given the subject their particular attention, the number has been gradually swelled, until at the present time it amounts to upwards of three hundred and twenty distinct and well-defined species. This result is in a great measure owing to the energetic exertions of John Gould of London and Charles Lucian Bonaparte, Prince of Canino, whose collectors have distributed themselves throughout the continents of North and South America, making search among unexplored regions for new species.

The warm and ever-glowing countries of the Tropics seem to be the most favorite resort of this lovely tribe, before the brilliant fire of whose sparkling plumage the gorgeous colorings of all other feathered races grow dull. There, revelling in the undying

Ruby-throated Humming Bird.

verdure of a perpetual Summer, these gems of the forest sport their charms amid the sweets of a thousand flowers.

Although by far the largest number of species of the Humming Birds are found in the West Indies, the Brazils, and those countries which lie adjacent to the Equator, yet these are by no means the limits to which they are confined : they enjoy probably the most extensive range of country, and experience the greatest variety of climate, of any known family. The continents of North and South America, from Nootka Sound on the northwest and Canada on the northeast, to Terra del Fuego on the south, can alone be given as the limits of their migrations. The beautiful and lovely little bird discovered by Captain Cook on the borders of Nootka Sound, and which inhabits the whole northwest coast, is a lonely representative of the genus in the ornithology of those parts; while in Canada and the United States, the little Ruby-throated Humming Bird is, during Summer, a welcome delegate of the tribe. And although it does not, like the Wood Thrush, sit and regale us by its melodious song, yet we are none the less attracted by its tiny form, its activity and gracefulness as it flits from flower to flower, and gaze with admiration upon the sparkling of its jewelled breast.

As we advance farther south the species become more numerous; in Mexico and Guatemala we find upwards of thirty or forty species; while in the West Indies and the vast expanse of Central America, there are comparatively few members of the family

Black-headed Humming Bird.

that may not be found at some season of the year. On the lone island of Juan Fernandez, in the vicinity of the very cave which tradition has dedicated to the memory of the renowned Robinson Crusoe, two elegant species have been obtained. In the Andes, whose lofty summits are capped with eternal snows; in the deep recesses of the craters of extinct volcanoes; and where Chimborazi and Cotopaxi poured forth their streams of liquid fire, — there these little jewelled bands, with untiring wing, suck the sweet nectar from some favorite flower, or with the velocity of thought flash like meteors across the vision, in pursuit of their prey.

The traveller who has visited the haunts of these birds, can alone possess an adequate idea of their surpassing loveliness. As they seldom live long in confinement, almost the only impression we can form of them is gained from the descriptions of those who have observed their habits in their native woods, and from the examination of the stuffed skins in our cabinets. The varieties of form, size, and color are so many, and the general development of the organs is so various, that in viewing a collection of these lovely creatures, one cannot but wonder at so wide a difference between them, while a general resemblance is constantly preserved.

In the island of Jamaica, and peculiar to that locality, is found a species familiarly known by the name of Polytmus, or Black-headed Humming Bird, — having two of the tail feathers lengthened to a degree quite disproportionate, being more than twice the length of the body; while in the Andes of Bogota there exists a variety (Sword-bill) with the bill protruded to such an astonishing extent, as almost to make one laugh at the magnitude of the supposed deformity. From Brazil and Guiana we receive specimens having crests on the head, and lateral tufts on the neck, capable of being raised or depressed at pleasure, and which, when fully expanded, give the bird the appearance of being possessed of two

14

Sword-bill Humming Bird.

pair of wings. Of these, the Chestnut-tufted Co-
quette is the most beautiful. Others again with
crests of various forms and dimensions; some, as in
the Delalande Humming Bird, whose crest when
elevated measures nearly as long as the body of the
bird.

But the most striking difference is in the various
forms and peculiar development of the tail. The
Polytmus, with its long, dangling plumes, has already

been noticed. In the Racket-tail we have a tail deeply forked, with the two outer quills entirely bare of webbing in the centre, for about one-third their length, and at the outer extremities expanding suddenly into a broad spathe, somewhat in the shape of a battledore. In the gorgeous Comet-tail the tail is forked, and composed of broad feathers, the outer pair about four and a half inches in length, all tipped with black, and glowing with a radiant lustre like highly polished brass, with a considerable tinge of red, which has given it with some the significant title of Flame-tail. A number of other species might be mentioned to show the vast variety of forms assumed by this important appendage, which adds to each species a peculiar grace, and no doubt exerts a considerable influence in regulating its motions upon the wing; but the limits of this article will not admit of more.

The peculiar beauty of the Humming Bird consists in the metallic lustre of its coloring; and when seen in a strong light, some parts of the plumage exhibit a surface of the most exquisite polish, glittering with all the brilliancy of the ruby, the fiery lustre of the topaz, and the soft sparkling of the sapphire, the emerald, and the amethyst. Their voice consists mostly of a low twitter or chirp, although it is asserted that some species indulge in a low but not unmusical warble. Thus we see that Nature distributes her gifts with an equal hand; for, while to these little creatures she has given a plumage of the most unrivalled splendor, covering their feathers with burnished gold, and tinging them with the ever-changing hues of the most glittering gems, — upon others, arrayed in a plainer dress, she has bestowed that peculiarly fascinating and delightful charm, a voice that rings through the woodlands like a heaven-born melody.

It has been observed that the Humming Birds seldom live long in confinement; and although they have been kept during a period of several weeks, yet they generally languish and die in a much shorter space of time. A creature so evidently formed for continued activity, whose very food is taken upon the wing, would naturally prove difficult to domesti-cate; and the impossibility of supplying it with its natural food, would at once suggest the uncertainty of success. The Polytmus has been known in sev-eral instances to live in an apartment sufficiently large to allow of free exercise; and by being constantly supplied with fresh flowers and a syrup prepared for the purpose, has been kept alive for a few weeks; but the almost entire absence of the minute insects which constitute the principal part of their food, rendered them so feeble and emaciated as soon to cause their death from actual starvation. When first caught and placed in confinement, they mostly pine away, and die in a few days of fright or grief. Sometimes, in fits of desperation, they beat themselves about and butt their little heads against the sides of the cage, and soon fall down exhausted and die.

In the manner of constructing their nests, the Humming Birds differ almost as widely as in their

forms and colorings. In some species it is hung in
the most graceful manner from the tendrils of some
twining creeper, whose luxuriant bowers of fragrant
bloom supply them with abundant food and protection
from the weather. Some are supported by the slender stalks of a rampant shrub, while others are perched
beneath the jutting point of some rock o'ergrown
with ferns and flowers, or built upon the horizontal
branch of some moss-covered tree. The beautiful
Delalande Humming Bird constructs a neat little
nest in the form of an inverted cone, made of moss,
lichens, fibrous roots, spiders' webs, and the involu-
cres of plants, suspended from the slender stems of a
species of bamboo, and almost entirely imbedded in
its foliage. The little Ruby-throat of the United
States, the only species which is familiar to us, generally builds upon the strong branch of some old
tree, and so assimilates the outside of the nest with
the mossy covering of the bark, as to make it difficult to be discovered, except by accident or by dili-
gent search. The principal materials used in the
construction of the nests are fine grass, fibrous roots,
bark, spiders' webs, feathers, wool, hair, moss, and
lichens, each selecting such of them as are best
adapted to its wants, or most easily procured; and in
most, if not all cases, the interior is lined with the
soft down or pubescence gathered from various plants.

The following interesting account, given by a resident of Jamaica, of the manners of the Polytmus,
as having come under his own observation, is taken

14 *   L

from Martin's "Humming Birds of Gould's Collection :"

"In the latter part of February a friend showed me a nest of this species, in a singular situation, but which I afterward found to be quite in accordance with its usual habits. It was at Bognie, situated on the Bluefield Mountain. About a quarter of a mile within the woods, a blind path, choked up with bushes, descends suddenly beneath an overhanging rock of limestone, the face of which presents large projections and hanging points, encrusted with a rough tuberculous sort of stalactite. At one corner of the bottom there is a cavern, in which a tub is fixed, to receive water of great purity, which perpetually drips from the roof, and which in the dry season is a most valuable resource. Beyond this, which is very obscure, the eye penetrates to a larger area, deeper still, which receives light from some other communication with the air. Round the projections and groins of the front, the roots of the trees above have entwined, and to a fibre of one of these, hanging down, not thicker than a whip-cord, was suspended a Humming Bird's nest, containing two eggs. It seemed to be composed wholly of moss, was thick, and attached to the rootlet by the side. One of the eggs was broken. I did not disturb it, but after three weeks visited it again. It had apparently been handled by some curious child, for both eggs were broken and the nest evidently deserted. While I lingered in this romantic place, picking up some of the land shells which were scattered among the rocks, suddenly I heard the

whirr of a Humming Bird, and looking up saw a
female Polytmus hovering opposite the nest with a
mass of silk-cotton in her beak. Deterred by the
sight of me, she presently retired to a twig a few
paces distant, on which she sat. I immediately sunk
down among the rocks as gently as possible, and remained perfectly still. In a few seconds she came
again, and after hovering a moment disappeared behind one of the projections, whence in a few seconds
she emerged again, and flew off. I then examined
the place, and found to my delight a new nest in all
respects like the old one, but unfinished, affixed to
another twig not a yard from it. I again sat down
among the stones in front, where I could see the nest,
not-concealing myself, but remaining motionless, waiting for the bird's reappearance. I had not to wait
long : a loud whirr, and there she was, suspended in
the air before the nest. She soon espied me, and
came within a foot of my eyes, hovering just in front
of my face. I remained still, however, when I heard
the whirring of another just above me, perhaps the
mate; but I durst not look toward him, lest the turning of my head should frighten the female. In a
minute or two the other was gone, and she alighted
again on the twig, where she sat some little time
preening her feathers, and apparently clearing her
mouth from the cotton fibres, for she now and then
swiftly projected the tongue an inch and a half from
the beak, continuing the same curve as that of the
beak. When she arose it was to perform a very in-

teresting action; for she flew to the face of the rock, which was thickly clothed with soft dry moss, and hovering on the wing as if before a flower, began to pluck the moss until she had a large bunch of it in her beak. Then I saw her fly to the nest, and having seated herself in it, proceed to place the new materials, pressing and arranging and interweaving the whole with her beak, while she fashioned the cup-like form of the interior by the pressure of her white breast, moving round and round as she sat. My presence appeared to be no hindrance to her proceedings, although only a few feet distant; at length she left the place, and I left also."

In all the species, as far as has yet been ascertained, the female deposits but two eggs, which are beautifully white, or slightly tinged with yellow; the period of incubation varies from ten to about sixteen days; the young, when hatched, are quite naked and blind, but soon become covered with feathers, and in about three weeks are able to take care of themselves and leave the nest, becoming in a short time as active on the wing as their parents, from whom they can only be distinguished by their plumage.

The fact that the food of these birds consists mostly of insects, has been well established both by observation and experiment; the few individuals which have lived in confinement have been seen eagerly catching such as have chanced to be in the apartment which they occupied; while the quick snapping of the bill, similar to that of the Fly-catchers, distinctly heard when darting through the air, at once indicates the nature of its sustenance. For this reason they often frequent the borders of streams; and are seen skimming over the surface of ponds of water, where a minute insect life is most abundant. The bills also of many species are provided with seratures, to enable them more certainly to secure their prey. The corollas of many large tubular flowers are infested by microscopic insects, which undoubtedly attract the birds, as well as the sweet nectar contained in the cup below; and to obtain which they are furnished with a tongue formed like that of the Woodpeckers, divided into two tubes which run throughout its entire length, and is capable of being protruded to a considerable distance from the point of the beak, thus serving the purpose of a pump to draw up the honey from the deep recesses of the flower, while it is also used to collect the insects from the corolla.

In most species of Humming Birds there is a wide difference noticeable in the plumage of the males and the females, the latter being rarely if ever clothed with the rich metallic hues of the former. In a few instances where the coloring of both sexes is plain, no difference is apparent. The young birds do not generally attain their full livery until the second or third year; they make their first appearance in the sombre garb of the female, which gradually changes with each successive moulting until maturity.

The structure of the scale-like feathers which

adorn various parts of their bodies is very peculiar, presenting as they do a beautifully burnished surface, glittering with intense brilliancy, and tinged with the most exquisite shades of green, gold, crimson, or black.

The Ruby Topaz Humming Bird, when viewed directly in front, has a gorget of the most fiery orange; but alter the angle at which the light strikes it, and we have a surface of emerald green, which, by still another change in position, is converted into velvety black. This changeableness is due to the construction of the feathers, for, upon close examination, we find each composed of a multitude of facets, which are so arranged as to present various angles to the falling rays, and thus absorb or reflect the different colors, according to the position in which they are held.

The surpassing beauty, the swiftness of flight, and the apparent intelligence of these winged gems, cannot fail to attract and rivet the attention of the most listless observer; darting from blossom to blossom, poising themselves as by magic, in mid air, upon viewless wings, now gently dipping their radiant bosoms into the deep recesses of the gayest corollas, and now resting like little fairies upon some delicate twig or tendril to preen their ruffled plumes, they must ever be to the reflective mind fit objects of wonder and admiration.

There are seven species which have been found within the limits of the United States. The Ruby-throat, abundant almost everywhere in Summer; the

Nootka Sound Hummer, inhabiting Oregon and the Northwest Coast; the Anna Humming Bird; the Purple-throated Humming Bird; and Costé's Humming Bird, found in California and Mexico; the Broad-tailed Flame Bearer, from Texas and Mexico; and the Mango Humming Bird, a single specimen of which was captured upon one of the small islands or keys at the southern extremity of Florida.

CHAPTER IX.

INSESSORES: *ACCIPITRES.*

WALK TO THE FIELDS — HABITS OF DIFFERENT BIRDS — TUR-
KEY BUZZARD — VULTURE — CONDOR — EAGLE — HAWK —
FALCON — KITE — HAWK OWL — AND THE OWL.

IF we look over the wide extent of our country,
washed by the bright waves of the Atlantic on the
one side, and by the blue waters of the Pacific on
the other, and stretching from the cold icy regions
of Hudson's Bay to the far-off boundaries of Texas
and California, we shall observe that its surface is
not only diversified with a charming variety of mountains and valleys, hills and dales, table-lands and
prairies, but that each region is tenanted by an animated life in many respects peculiarly its own. This
is particularly noticeable with reference to birds ;
and although many species seem to enjoy a wide
range, extending during their migrations almost from
the extreme north to the extreme south, yet it will
be found that the summer haunts of most are generally restricted to certain localities, beyond which they
are seldom known to build their nest and rear their
young. This peculiarity will become more apparent
as we proceed with our description of some of the
most prominent species.

We will now invite our readers to accompany us

Snowy Owl.

(169)

15

into the fields and woods, far from the noise of city
life, and where no sound is heard but the ceaseless
voice of Nature. Here we shall see the birds in all
their native beauty, not as we see the stuffed mum-
mies in our cabinets, but as free tenants of the air,
enjoying all the life and liberty in which they were
created. It is a warm, bright morning of Summer;
the sultry air teems with the fragrant odors of the
hay-fields; the sweet warblers which early sang their
notes from the neighboring grove have retired to the
deep and cooling shelter of the forest. We seek the
shade of some wide-spreading oak, where we may sit
down and observe what is passing around us. If we
turn our eyes upward, we will probably see four or
five dark-looking objects, apparently like crows, sailing in easy circles, or floating about in graceful curves,
sometimes dashing off with impetuous velocity, or
mounting high in the air, until almost lost to view,
their varied motions being performed without any
further apparent effort of the wings than a few flaps.
These are the Turkey Buzzards, and if one of them
should pass before us upon the ground, we would
scarcely suspect so awkward, unsightly, heavy and
inanimate a looking object, could be so free and
graceful upon the wing; and if we should see him
thrust his head and neck into the mangled corpse
of some poor old horse which had just fallen a prey
to the stroke of death, we should be still more dis-
gusted with his unmannerly behavior. But however justly we may censure him for his uncouth appearance and his filthy habits, he is nevertheless one

of our best friends.   In the warm cities of the South, (for it is here that these birds are most abundant), troops of them, in company with the Black Vultures, may be found almost daily performing the office of scavengers.   They are to be seen walking or flying about the streets, frequenting the markets or sham-bles, and greedily snatching up the pieces of flesh which are thrown away by the butchers, and even attempting, when opportunity offers, to help them-selves from the benches where meat is exposed for sale ; thus the air is, in great measure, kept free from the foul effluvia which would otherwise be created by the accumulation of such substances.   They will also follow the carcass of a horse or cow as it is dragged through the streets, and upon its being deposited in the suburbs, will even dispute possession with the dogs which assemble to assist in devouring it ; but should Eagles make their appearance on such occa-sions, the Vultures retire, and patiently wait until their second turn comes, when they immediately commence again in all the hurry of a keen appetite, and seldom stop until the whole is consumed.

The California Vulture is another species similar in its habits and appearance, although much larger, it being the largest bird known to exist north of the isthmus of Darien, almost equalling the far-famed Condor of the Andes, to which it is closely allied.*

---

* The Condor, being a large and powerful bird, is, even under unfavorable circumstances, almost a match for a full-grown man.   Captain Head relates the following anecdote

Vultures can perceive the existence of carrion at a very great distance. Some authors have supposed that it was owing to the sense of smell being very acute ; but it appears to be by no means certain that the olfactory nerve, which in mammalia is the organ of smell, does in birds perform that function. The Vultures, as well as many other birds, possess an

---

of a contest between a strong English miner from Cornwall and one of these gormandizers after a full meal :

"The man, when riding along the plains, saw several Condors, and guessing that they were attracted by the body of some dead animal, rode up, and found a numerous flock around the carcase of a horse. One of the largest was standing with one foot on the ground, and the other in the horse's body, exhibiting a singular force of muscular power, as he lifted the flesh and tore off great pieces, sometimes shaking his head and pulling with his beak, or sometimes pushing with his leg. As the man approached, one of them, which appeared to be gorged, rose up, and flew about fifty yards off, when it alighted, and he rode up to it, and then jumping down, seized the bird by the neck. The contest was severe, and never probably was such a battle seen before, as a Cornish miner and a Condor. The man declared he never had had such a trial of strength in his life, that he put his knee upon the bird's breast, and tried with all his might to twist his neck, but that the Condor, objecting to this, struggled most violently, and he fully expected that several other birds, which were flying over him, would take part against him, and assist their companion. At length, however, the man succeeded, as he supposed, and carrying off the pinion quills in triumph, left the bird for dead. But so tenacious are they of life, and so difficult to kill, that another horseman, who passed the spot some time after, found it still living and struggling."

15 *

extraordinary development of the nasal organs, but
for what purpose it is designed is not fully known.
From the earliest ages, the powers of vision of these
birds have been almost proverbial, and as they seem
to be constantly on the look-out for some object with
which to gratify their voracious appetites, it is more
than probable that their quickness of sight, rather
than the sense of smell, assists them in discovering
their food.   Wilson, the American ornithologist,
speaks of having counted two hundred and thirty-
seven black Vultures, which had collected for the
purpose of devouring the carcase of a horse; and
from his description, we should suppose that nothing
was left but the naked skeleton long before the least
effluvia could have escaped from the body to attract
them.*

Sometimes in the midst of a troop of Vultures may
be recognized the white head and white tail of the

---

* In travelling over the wide deserts of Africa, where
there is not a blade of grass to tempt a living bird or animal, and therefore no inducement for birds of prey to scour
the wilderness in search of game, should a camel or other
beast of burden drop under its load in the train of a caravan,
in less than half an hour there will be seen high in the air
a number of the smallest specks moving slowly round in
circles, and gradually growing larger and larger as they
descend in spiral windings toward the earth. These are the
Vultures, but whence they come, or by what sign or call
they are collected from such a vast height, is mysterious;
though it is quite possible that it is in consequence of both
the senses being possessed of an acuteness of which we can
hardly form any conception.

Bald Eagle, the rest of his plumage being quite similar to that of his less dignified companions, but from whom he may be readily distinguished by the greater ease of his motions, as well as his more majestic appearance. While the Turkey Buzzard sails in contracted circles, or swims off in a wide curve, the Eagle, as if conscious of his superiority, floats upon his unmoving wing as though he would compass in one vast sweep the broad expanse of Heaven; or sometimes, when at his greatest altitude, hardly appearing as more than a black speck in the dim distance, he will fold his wings and descend with the velocity of thought toward the earth, when suddenly unfurling his broad pinions, he checks his downward course, and glides off like an arrow to a distant quarter.

The Eagles often resort to stratagem to secure their prey, being well aware that ducks, and other waterfowl on which they feed, can readily elude their grasp by diving beneath the water and again appearing above the surface at some distance. To meet this difficulty, they will hunt in pairs; and having discovered the object of their search, will ascend into the air in opposite directions until they have reached a considerable height, when one of them immediately glides with great swiftness toward the place where the bird is engaged quietly seeking its food; the latter, observing his intentions, dives the moment before he reaches the spot, but upon again rising to the surface he is met by the second Eagle, whose keen vision may have traced his course under the

water, and who has descended from his elevation just
in time to force the poor bird again to take refuge
beneath the water almost before he has taken breath;
and thus by repeated attacks the duck becomes wea-
ried, and swims for the shore, where he is easily captured by the Eagles, who divide the dainty morsel
between them.

Fish also constitutes a considerable portion of the
food of the Bald Eagle, and to the vicinity of the
sea or other large bodies of water they often resort
for the purpose of obtaining it.   Here one may sometimes be seen "fishing," as the boys say, " upon his
own hook," but much more frequently does he supply himself and young with food by robbing the in-
dustrious Fish Hawk of the fruits of his honest toil.
The scene thus enacted is often of a very interesting
and exciting character, and is thus graphically described by Wilson : " Elevated on the high dead limb
of some gigantic tree that commands a wide view of
the neighboring shore and ocean, the Eagle seems
calmly to contemplate the motions of the various
feathered tribes that pursue their busy avocations
below, — the snow-white Gulls slowly winnowing the
air ; the busy Fringæ coursing along the sands; trains
of Ducks streaming over the surface; silent and watchful Cranes, intent and wading; clamorous Crows; and
all the winged multitudes that subsist by the bounty
of this vast liquid magazine of Nature.   High over
all these hovers one, whose action instantly arrests
his whole attention.   By his wide curvature of wing,
and sudden suspension in air, he knows him to be

the Fish Hawk, settling over some devoted victim of the deep. His eye kindles at the sight, and, balancing himself, with half-opened wings, on the branch, he watches the result. Down, rapid as an arrow from Heaven, descends the distant object of his attention, the roar of its wings reaching the ear as it disappears in the deep, making the surges foam around. At this moment the eager looks of the Eagle are all ardor; and, levelling his neck for flight, he sees the Fish Hawk once more emerge, struggling with his prey, and mounting in the air with screams of exultation. These are the signal for our hero, who, launching into the air, instantly gives chase, and soon gains on the Fish Hawk; each exerts his utmost to mount above the other, displaying in these rencontres the most elegant and sublime aërial evolutions. The unencumbered Eagle rapidly advances, and is just on the point of reaching his opponent, when, with a sudden scream, probably of despair and honest execration, the latter drops his fish; the Eagle, poising himself for a moment, as if to take a more certain aim, descends like a whirlwind, snatches it in his grasp ere it reaches the water, and bears his ill-gotten booty silently away to the woods."

This cowardly and selfish behavior of the Eagle would seem to unfit him to be the national emblem of a people devoted to freedom, and who glory in the unmolested enjoyment of their rights. Dr. Franklin deeply regretted that it had been chosen as the representative of our country, but however appropriate or inappropriate the comparison may be, there is no

good reason for our following its example in idly
watching the labors of the poor slave, and then robbing him of a part of the fruits of his toil.

The most noble representatives of this family are
the Golden Eagle and the Washington Eagle, both
natives of America, and the former of many parts
of Europe and Asia.

The Golden Eagle is a large and powerful bird,
noble and majestic in appearance. Its food consists
principally of lambs, fawns, rabbits, turkeys, ducks,
and other large birds. In capturing its prey it does
not manifest the same agility as the Bald Eagle in
pursuing and seizing it upon the wing, but it is
obliged to descend from a considerable height upon
it to insure success. The keenness of its vision,
however, enables it to discern at a great distance the
objects of its desire, upon which it generally falls
with the swiftness of a meteor, and with an unerring
and deadly aim. The feathers of this Eagle are
much sought after by the Indians of North America,
as an ornament of their dress; and so highly are they
prized, that it is said a warrior will often exchange a
valuable horse for the tail feathers of a single bird.*

---

* Eagles being possessed of both strength and courage,
will, under some circumstances, especially when pressed by
hunger, openly attack the human species; and numerous
well-authenticated accounts are on record of young children having been carried away and devoured by them.
Bishop Heber, in his travels in India, passed through a
mountainous district where sad complaints were made of
their carrying off infant children; and some years ago a
traveller in the Alps observed suspended from a jutting crag

Next in size and importance to the Eagles come the Hawks and Falcons, of which the varieties are numerous. They all possess great similarity in their formation and habits, mostly pursuing their prey upon the wing, securing its capture by the vigor and rapidity of their flight.

The Sparrow Hawk, a neat and very active bird, rather less in size than a pigeon, is a frequent visitor to the farm-house and barn-yard, where it sits perched erect upon a fence-stake, watching intently for the approach of some unlucky mouse or mole, or even for beetles or grasshoppers, upon which it pounces with great quickness, and immediately returns to its stand to devour it. When changing its position it flies low until within a few yards of the spot upon which it wishes to settle, when it suddenly

---

the tattered remains of a child's clothing, who had been carried away from the valley below by the Lammergeier, or Bearded Vulture.

A large Eagle some years ago made an attack upon a little boy about seven years of age, residing near the city of New York, who, with a younger brother, was amusing himself with attempting to reap, during the absence of their parents. The bird sailed slowly over them, and with a sudden swoop endeavored to seize the child, but luckily missed him. He then alighted at a short distance for a few moments, when he again renewed the attempt. The brave little fellow at once struck at his assailant with the sickle which he happened to have in his hand, and so resolutely was the blow given, that entering under the left wing it passed between the ribs, and penetrating the liver, proved fatal. The bird's stomach was found to be entirely empty, which may in some degree account for so unusual an attack.

rises with an easy curve and alights with the utmost grace, closing its wings with the rapidity of thought. Sometimes a Sparrow or Finch crosses its pathway, when the little Hawk, all anxiety to secure so great a prize, at once gives chase, and soon overtaking it, bears it off to share the dainty morsel with its mate and young. The Sparrow Hawk is capable of being domesticated and rendered quite companionable. Audubon gives the following description of a young bird which he kept for some time : "I once found a young male that had dropped from the nest before it was able to fly. Its cries for food attracted my notice, and I discovered it lying near a log. I took it home, named it 'Nero,' and provided it with small birds, at which it would scramble fiercely, although yet unable to tear their flesh, in which I assisted it. In a few weeks it grew very beautiful, and became so voracious, requiring a great number of birds daily, that I turned it out to see how it would shift for itself. This proved a gratification to both of us. It soon hunted for grasshoppers and other insects, and on returning from my walks, I now and then threw a dead bird high in the air, which it never failed to perceive from its stand, and toward which it launched with such quickness as sometimes to catch it before it fell to the ground. The little fellow attracted the notice of his brothers, brought up hard by, who, accompanied by their parents, at first gave it chase, and forced it to take refuge behind one of the window-shutters, where it usually passed the night; but soon became gentler toward it, as if forgiving its desertion.

Upper fig.—Sparrow Hawk. *Lower fig.*—Pigeon Hawk.

16

My bird was fastidious in the choice of food, would not touch a Woodpecker, however fresh, and as he grew older refused to eat birds that were in the least tainted. To the last he continued kind to me, and never failed to return at night to his favorite roost behind the window-shutter. His courageous disposition often amused the family, as he would sail off from his stand and fall on the back of a tame duck, which, setting up a loud 'quack,' would waddle off in great alarm, with the Hawk sticking to her. But, as has often happened to adventurers of similar spirit, his audacity cost him his life. A hen and her brood chanced to attract his notice, and he flew to secure one of the chickens, but met one whose parental affection inspired her with a courage greater than his own. The conflict, which was severe, ended the adventures of poor Nero."

The Duck Hawk is probably the swiftest-winged Hawk with which we are acquainted. When pursuing its prey it moves with astonishing rapidity, following it in all its turnings and dodgings through the air until within a few feet of it, when it protrudes its talons, and closing its wings for a moment, rushes upon it, and if not too heavy, bears it off to the earth. He pursues the Ducks and Water Hens with such quickness as often to snatch them from the water before they could dive beneath it; and with the most daring assurance will sometimes come at the report of a gun and carry off the prize almost from under the nose of the sportsman who has killed it.

The Peregrine Falcon, which is a native of Europe,

representing there the Duck Hawk of America, appears to have been the favorite Hawk among the falconers of the olden time. In the early part of European history mention is frequently made of the sport of hawking, and it was then considered as a recreation of such a dignified character, that it was placed by laws beyond the power of any but the nobility to engage in it. The various nobles vied with each other in the superiority and numbers of their Falcons, and the life of a serf is said to have been esteemed of less value in the eyes of a Norman Baron than that of his favorite Hawk.

Hawk Owl.

To the Hawk family also belong the Kite, the Swallow - tailed Hawk, the Pigeon Hawk, the Sharp-shinned Hawk, and the Red - shouldered and Red - tailed Buzzards, all of which are more or less abundant in the various sections of the country.

Next to these, as a connecting link between the Hawks and Owls, we have the Hawk Owl, which appears to be only an occasional visitor south of the St. Lawrence river. In the vicinity of Hudson's Bay it is quite abundant, and is also found in Denmark, Sweden, and Siberia. It strongly resembles the Hawks in the general form of the body, the narrowness of the face, and the length of the tail; but the radiating feathers around the eyes and bill, as well as the form of the legs and feet, at once distinguish it as an Owl. It is said to be a bold and active species, possessing many of the manners of the Hawk, preying by day, and often following the sportsman and carrying off the game as soon as shot.

With the general appearance of the Owl it is presumed that most of our readers are acquainted. A large head, with a broad flat face, huge eyes surrounded with fine feathers, which radiate in all directions, and almost conceal its small, hooked bill; the head sometimes surmounted with two fierce-looking horns which project sideways from above the eyes; these form some of the most prominent features of this peculiar family.

With the Owl has generally been associated the habit of prowling about at night, and committing all kinds of depredations upon its sleeping fellow-creatures, and occasionally scaring some dreamy slumberer by perching upon his window-shutter, and interspersing his visions with a wild and unearthly laugh. How often has this innocent note of the poor little Owl been made the foundation of senseless stories about ghosts and other appearances whose existence is not only contrary to Nature, but utterly impossible!

It is observed that in most species of Owls the wing is formed of soft and downy feathers, in conse-

16 *

quence of which its flight is noiseless, and it glides
through the still air and pounces upon its victims
without awaking them, until too late to elude its
grasp. But there are some varieties in which this
formation is not so noticeable; they are generally
found seeking their food by day, and possessing all
the activity and vigor common to other diurnal birds
of prey.

Of these we will mention the Great White or
Snowy Owl, inhabiting the same district of country
as the Hawk Owl, and several smaller varieties which
are active upon the wing in broad daylight. The
Snowy Owl is only a winter resident in the United
States, retiring during the Summer to the Arctic regions. It is, as its name indicates, of a beautiful
snowy whiteness, sometimes, especially in Summer,
marked with spots of brown. It feeds on various
small quadrupeds, on Ducks and other water-fowl,
and frequents the margins of rivers and creeks for
the purpose of fishing. They will sometimes, when
pressed for food, watch at a hole in the ice for the
fish to pass, when they will catch them in the most
dexterous manner. Audubon gives the following
interesting account of this peculiar habit of the
bird: "At the break of day, one morning, when I
lay hidden in a pile of drift logs at that place (the
Falls of the Ohio, at Louisville, Kentucky,) waiting
for a shot at some wild geese, I had an opportunity
of seeing this Owl secure fish in the following manner:—While watching for their prey on the borders
of the 'pots,' they invariably lay flat on the rock,

with the body placed lengthwise along the border of the hole, the head also laid down, but turned toward the water. One might have supposed the bird sound asleep, as it would remain in the same position until a good opportunity of securing a fish occurred, which, I believe, was never missed; for, as the latter unwittingly rose to the surface, near the edge, that instant the Owl thrust out the foot next the water, and, with the quickness of lightning, seized it, and drew it out. The Owl then removed to the distance of a few yards, devoured its prey, and returned to the same hole; or, if it had not perceived any more fish, flew only a few yards over the many 'pots' there, marked one, and alighted at a little distance from it. It then squatted, moved slowly toward the edge, and lay as before, waiting for an opportunity."

The Night Owls, with which we are most familiar, are the Great Horned, the Long-eared, the Short-eared, and the Little Screech Owls. The latter is the most abundant species, and there is scarcely any section of the Eastern and Middle States where it is not found. Its melancholy notes are heard around the doors of our farm-houses, as it sits perched upon a neighboring tree. Its song, if song it may be called, resembles somewhat the syllables, "Who-o-o-o-o-oo-oo!" uttered through the nose tremulously, and sometimes conveys the impression that they proceed from a child in distress. These notes are most frequently heard during the latter part of Winter; and this being the mating season, the male bird is particularly attentive to the object of his affections, strutting about her in grotesque attitudes, and occa-

Screech Owl.

sionally saluting her with a nod or a bow, awkward enough to make one laugh.

In the vicinity of the Rocky Mountains a curious species of Owl is found, called the Burrowing Owl; it inhabits the deserted holes of the Marmots or Prairie Dogs, which are so abundant as sometimes to cover many acres of ground with their villages. In localities where these holes do not exist, the Owl is said to make a burrow for itself, at the bottom of which it lays its eggs. They appear to live on friendly terms with the Marmots, but never, as has been supposed, is the same burrow inhabited by both; the Owl always selecting for itself one where it may retain undisputed possession. Their habits are strictly diurnal, and they feed upon grasshoppers, crickets, and perhaps on field-mice. The nest is composed of fine grass, and placed at the extremity of the hole, where the bird deposits four pale white eggs, about the size of those of a pigeon.

CHAPTER X.

INSESSORES: *PULLASTRÆ.*
CURSORES: *GALLINÆ.*

PIGEONS — GREAT FLIGHT OF PIGEONS, BY "AUDUBON" —
TURTLE DOVE — WILD TURKEY — AMERICAN AND GAMBLE'S
PARTRIDGE — CANADA, RUFFLED, AND PINNATED GROUSE
— PTARMIGAN.

THE natural division of Birds called Pullastræ
embraces the Doves and Pigeons, the Australian
Brush Turkey, the extinct Dodo, etc.

Of all the different members belonging to these
several groups, by far the most interesting is the
Passenger, or common Wild Pigeon.　It is possessed
of some of the most singular habits which we have
yet had occasion to notice in any bird.　It is gifted
with the most astonishing powers of flight, both as
respects speed and continuance, one mile in a minute
being considered as the average rate at which it travels, and this often for many hours together.　But
the most remarkable characteristic of these curious
and interesting birds, is their habit of congregating
together at all seasons of the year, and in such numbers as we believe have no parallel among all the
feathered tribes of the earth.　During the period of
incubation their nests will occupy almost every available spot in a tract of woodland many miles in ex-

tent. In some instances they are so crowded upon
the branches, as to cause them to give way; and
when the young are fully fledged, and the place
finally deserted, so great has been the havoc and destruction they have caused, that what was before a
flourishing forest is converted into a wilderness of
dismantled trunks, every tree being as completely
destroyed as if girdled, and the whole ground covered with their excrements to the depth of several
inches.

But it is during their migrations that they assemble in the most astonishing multitudes. These migrations are performed only for the purpose of obtaining food, and are not influenced by any changes
in temperature, or the desire to seek a more genial
climate. Such countless thousands of hungry birds
must of necessity soon deprive a large tract of land
of all its available resources; hence the necessity of
their frequently changing their position.

Audubon, speaking of one of these companies,
says: "In passing over the Barrens, a few miles
beyond Hardinsburg, I observed the Pigeons flying
from northeast to southwest, in greater numbers than
I thought I had ever seen them before; and feeling
an inclination to count the flocks that might pass
within the reach of my eye in one hour, I dismounted,
seated myself on an eminence, and began to mark
with my pencil, making a dot for every flock that
passed. In a short time finding the task which I
had undertaken impracticable, as the birds poured in
in countless multitudes, I rose, and counting the dots

American Partridge.

Gamble's Partridge.

then put down, found that 163 had been made in twenty-one minutes. I travelled on, and still met more the further I proceeded. The air was literally filled with Pigeons; the light of noonday was obscured as by an eclipse.

" Before sunset I reached Louisville, distant from Hardinsburg fifty-five miles. The Pigeons were still passing in undiminished numbers, and continued to do so for three days in succession."

They are very fond of acorns, beech-nuts, and the smaller fruits of the forest trees generally; and when they have discovered a spot where these abound in sufficient quantities to induce them to alight, they do so in the most graceful manner, wheeling around in circles, as though to discover if danger were near. When fairly settled, they commence scratching among the leaves for food, which they swallow with such haste as sometimes fairly to choke in the process. Parts of the flock are almost constantly changing their position, which gives it the appearance of being continually in motion.

It is a singular circumstance that the roosting-places of these birds should be at so great a distance from the spots where they feed, being sometimes as much as sixty or eighty miles apart. This is no doubt occasioned by their being compelled to change their feeding ground frequently, while they still return to the same nightly rendezvous.

One of these roosts is thus described by Audubon: " It was, as is always the case, in a portion of the forest where the trees were of great magnitude, and

17   N

where there was little underwood. I rode through it upwards of forty miles, and, crossing it in different parts, found its average breadth to be rather more than three miles. My first view of it was about a fortnight subsequent to . the period when they had made choice of it, and I arrived there nearly two hours before sunset. Few Pigeons were then to be seen, but a great number of persons, with horses and wagons, guns and ammunition, had already established encampments on the borders. Two farmers from the vicinity of Russelville, distant more than a hundred miles, had driven upwards of three hundred hogs to be fattened on the pigeons which were to be slaughtered. Here and there, the people employed in plucking and salting what had already been procured, were seen sitting in the midst of large piles of these birds. Many trees two feet in diameter, I observed, were broken off at no great distance from the ground; and the branches of many of the largest and tallest had given way, as if the forest had been swept by a tornado. Everything proved to me that the number of birds resorting to this part of the forest must be immense beyond conception. As the period of their arrival approached, their foes anxiously prepared to receive them. Some were furnished with iron pots containing sulphur, others with torches of pine-knots, many with poles, and the rest with guns. The sun was lost to our view, yet not a Pigeon had arrived. Everything was ready, and all eyes were gazing on the clear sky, which appeared in glimpses among the tall trees. Suddenly there burst forth a

general cry of 'Here they come!' The noise which
they made, though yet distant, reminded me of a
hard gale at sea, passing through the rigging of a
close-reefed vessel. As the birds arrived and passed
over me, I felt a current of air that surprised me.
Thousands were soon knocked down by the pole-men.
The birds continued to pour in. The fires were
lighted, and a 'magnificent, as well as wonderful and
almost terrifying, sight presented itself. The Pigeons, arriving by thousands, alighted everywhere one
above another, until solid masses were formed on the
branches all round. Here and there the perches
gave way under the weight, with a crash, and falling
to the ground, destroyed hundreds of the birds beneath, forcing down the dense groups with which
every stick was loaded. It was a scene of uproar
and confusion. I found it quite useless to speak, or
even to shout to those persons who were nearest to
me. Even the reports of the guns were seldom
heard, and I was made aware of the firing only by
seeing the shooters reloading.

"No one dared venture within the line of devas-
tation. The hogs had been penned up in due time,
the picking up of the dead and wounded being left
for the next morning's employment. The Pigeons
were constantly coming, and it was past midnight
before I perceived a decrease in the number of those
that arrived."

The Passenger Pigeon is quite abundant in almost
all parts of the Union, — roaming in wild and un-
controllable masses from one place to another, now

appearing in one section of country, and then quitting it for an absence of years. Its plumage, though plain, is beautifully varied on the neck and shoulders with glossy feathers, reflecting in different lights the resplendent colors of the rainbow.

The Carolina Dove is another very abundant species, being found in the breeding season in nearly every part of the Union. They do not, however, like the Passenger Pigeon, assemble in large flocks, seldom being known to congregate in greater numbers than two or three hundred together, and that only during the period of migration. So very common and familiar are these birds, that it is difficult to take a ride of many miles into the country without meeting with them along the road-side, always flying in pairs, keeping some distance ahead of your vehicle, and now and then alighting in the middle of the road to search for food or to dust themselves. Thus you may follow them for some distance, until they suddenly wheel off into an adjoining field or wood. Their flight is very swift,

Passenger Pigeon.

and when surprised the motion of the wings is so rapid as to produce a peculiar whistling sound. They are constant residents of the Middle and Southern States, and during the Winter become very tame and sociable, sometimes resorting to the barn-yard, and feeding in company with the poultry.

Through the Pigeons we pass readily from the Insessores to the Gallinæ. This order comprises the well-known Wild Turkey, the Partridges, the Grouse, Pheasant, Guinea-fowl, etc.

The Wild Turkey, once so abundant in that part of the country lying between the Alleghanies and the Mississippi river, appears now to have become quite a scarce, shy, and in some places an obsolete bird. Like the poor Red Man who once roamed unrestrained through the same trackless woods, the march of civilization has encroached upon its freedom. And as the Indian has folded his blanket and gradually retired before the irresistible step of the avaricious white man, to the plains of the Far West, so this noblest game of the forest has taken its flight from haunts where once the murderous gun was seldom heard to echo, to nestle among the secluded wilds west of the Mississippi. Straggling companies, however, still remain in the yet unsettled parts of Pennsylvania, New York, and several of the Western States, though only relics of what was formerly a numerous and powerful tribe.

The Wild Turkey, from its weight and bulky proportions, is essentially a terrestrial bird; its food consists of the fruits of forest trees, which it searches

17 *

for beneath the fallen leaves, and such berries and small fruits as are within the reach of its very limited flight.

In the early part of the Autumn the Turkeys collect in small companies, the gobblers by themselves, and the old hens with their troops of young, which are but about half grown. They then commence to move about in search of fallen acorns and other small nuts. They travel on foot except when their progress is intercepted by rivers, or when surprised and forced to take wing by an enemy. Audubon says: "When they come upon a river, they betake themselves to the highest eminences, and there often remain a whole day, or sometimes two, as if for consultation. During this time, the males are heard *gobbling*, calling, and making much ado, and are seen strutting about as if to raise their courage to a pitch befitting the emergency. Even the females and young assume something of the same pompous demeanor, spread out their tails, and run round each other, *purring* loudly, and performing extravagant leaps. At length, when the weather appears settled, and all around is quiet, the whole party mounts to the tops of the highest trees, whence, at a signal, consisting of a single *cluck*, given by a leader, the flock takes flight for the opposite shore. The old and fat birds easily get over, even should the river be a mile in breadth; but the younger and less robust frequently fall into the water, — not to be drowned, however, as might be imagined. They bring their wings close to their body, spread out their tail as a support, stretch forward their neck, and striking out their legs with great vigor, proceed rapidly toward the shore; on approaching which, should they find it too steep for landing, they cease their exertions for a few moments, float down the stream until they come to an accessible part, and by a violent effort generally extricate themselves from the water."

The plumage of the old males is very beautiful, being almost wholly of a rich golden bronze, while that part of the neck and head, which are mostly bare of feathers, and the loose skin of the throat, commonly called the wattle, are of different shades of blue, purple, and red. They lose most of these bright tints upon being domesticated, and after the second year can scarcely be distinguished from the common breeds.

The Partridge family, to which we next invite attention, has recently been increased in number by the addition of several very interesting and beautiful species. When Alexander Wilson wrote his Ornithology, his knowledge of this group was apparently confined to the one species which he describes. At a later date, when Audubon was instituting his inquiries among the birds of our Western Territories, he added three more, and still more recently three or four additional varieties have been discovered in the newly acquired territory of California and New Mexico. The plumage of all the species is plain, and the tints mostly sombre, but of such exquisite blendings as give them a high rank for beauty among the Birds of America.

On the cut at the head of this Chapter we have figured the common American Partridge and Gamble's Partridge. Of the former species perhaps most persons have some knowledge. To those who reside in the country it is by no means a stranger, especially in winter, when it often frequents the barn-yard to assist the fowls in appropriating their feed; while in summer, its clear loud call of " Bob White! Bob, Bob White!" is as well-known and familiar a voice as proceeds from the grove. There is something peculiarly pleasing in this love-note of the Partridge; the clearness and distinctness with which it is uttered is surprising, and the soft, mellow tones, as they come from a distance, are full of such sweetness that they quite inspire one with a love for the bird. It really consists of three syllables instead of two; the first being simply an aspiration, it is not heard at any great distance. Audubon makes the whole read, "Ah, Bob White!"

The nest of this bird is generally built at the foot of a tuft of grass or corn-stalks; it is slightly sunk below the surface of the ground, and is composed of grass so arranged as to form a sort of oven, with an opening at one side. The number of eggs deposited in one nest appears to vary from fifteen to twenty-four. The young leave the nest immediately upon being freed from the shell, and follow their mother in search of food, and nestling under her wings in the same manner as a brood of young chickens; they generally follow her until the succeeding Spring, when they are in full plumage, and capable of shifting for themselves.

Gamble's Partridge is an inhabitant of Texas, and was first discovered and introduced to notice by Dr. William Gamble, in 1841. For beauty of plumage it probably far surpasses any other species. The rich chestnut-colored feathers which cover the sides, the white markings upon the face and sides of the head, and the singular plumes with which the head is ornamented, give it a very sprightly and pleasing appearance.

General George A. M'Call, in his "Remarks on the habits of Birds met with in Western Texas, between San Antonio and the Rio Grande, and in New Mexico," speaking of this bird, says: "After losing sight of the Massena Partridge, I did not fall in with the present species until we reached the Limpia river, about 100 miles west of the Pecos.

"This beautiful bird, whose habits, in some respects, bear resemblance to the common Partridge, like that, seems to prefer a more genial and hospitable region. In this part of the country the Musquito Tree (Acacia Glandulosa) is more or less common; and the Musquito grass, and other plants bearing nutritious seeds, are abundant. Here, this Partridge increases rapidly in numbers, and becomes very fat; and, as I afterwards ascertained, is much disposed to seek the farms, if any be within reach, and to cultivate the acquaintance of man. About the Rancho of Mr. White, near El Paso, I found them very numerous; and here, in flocks of fifty or a hundred, they resort, morning and evening, to the barn-yard, and feed around the grain-stacks, in company with the poultry, where they receive their portion, as it is scattered amongst them by the hand of the owner." *

Of the Grouse family we number six species, only three of which are found to the eastward of the vicinity of the Mississippi river; these are the Canada Grouse, found only northward from the northern part of New York, and the Ruffed and Pinnated Grouse, which are very abundant, the former every-where north of Maryland, and the latter pretty generally distributed from Texas to Canada, more common in the west than to the eastward. These two species are, probably, next to the Wild Turkey, the finest game-birds which our Eastern States pro-duce. The markets of our cities are mostly well supplied with them during winter; the tenderness and delicacy of their flesh, and the fineness of its flavor, render it a great favorite with our epicures.

The Pinnated Grouse, or Prairie Hen, as it is called in the West, although clad in very plain col-ors, is nevertheless a handsome and stately bird, es-pecially when, during the love season, he struts about among his rivals with tail erect.and expanded, his head thrown backward, the lateral feathers on the neck spread to their utmost, the orange-colored drums beneath them swelled with air, and the wings stiffened and drooping in the manner of the Turkey

* Proceedings of Academy of Natural Sciences, Phila-delphia, 1851.

Gobbler. He is then in full dress; and his consequential attitudes and pompous manners give him quite an animated appearance. Companies of ten or twenty frequently assemble at daybreak, and perform these exciting manœuvres, which mostly result in fiercely contested battles, in which they attack each other in the manner of the common game-fowl.

Prairie Hen, or Pinnated Grouse.

The peculiar rolling or tooting sound which it generally makes before sunrise, although in the unsettled districts it is often heard from morning till night, is produced by inflating to their full extent the bladder-like appendages above the wings, and then throwing the head forward, forcing it through the throat in distinct rolling or undulating notes.

This sound, which is produced only by the male bird, can be heard at the distance of nearly half a mile.

The nest of this Grouse is usually placed in a tuft of tall prairie grass, or at the foot of a clump of low bushes. It is composed of dry leaves and grasses, neatly interwoven together. The female lays about twelve eggs, upon which she sits eighteen or nineteen days. The young leave the nest at once upon being hatched, and soon become quite strong and active. If a female and brood are surprised on the prairie, the latter immediately spread their little wings and scatter in all directions for a short distance, when they squat so close among the grass, that it is next to impossible to find them. In the Autumn several families club together and search for food in company until the Spring.

We can hardly pass from the order of Gallinæ without taking some notice of the Ptarmigan. There are several species of this beautiful and singular bird, which are occasionally found within the United States, but none of them are resident, their favorite haunts being among the icy regions of the north. The Willow Ptarmigan has been observed, during Winter, in the State of Maine and also in the Rocky Mountains. The White-tailed Ptarmigan is likewise a Rocky Mountain bird. They all, however, seem to prefer the more northern latitude of Hudson's Bay and the Fur Countries as a breeding-place, only leaving it for the south for a short time during the severity of Winter. One species, the Rock Ptarmigan, is found in the Rocky Mountains, in Greenland, Labrador, Norway, and Sweden. The plumage of this bird is very beautiful, and we have had occasion in another chapter to notice the singular change which it undergoes from one season to another. Dur-

Rock Ptarmigan.

ing Winter that of the male is of a snowy whiteness, with the exception of a broad band of black extending from the base of the bill to a short distance behind the eye, and the feathers of the tail, which are black. As Summer advances the white changes into a mixture of black, reddish-yellow, and white, beautifully varied, and marked with bars, spots, and bands of different shades. The female differs but little from the male, in Summer, the markings being perhaps a little less distinct.

18

CHAPTER XI.

CURSORES: *GRALLÆ.*

RAIL—WHOOPING CRANE—PLOVERS—SAND PIPERS—KILDEER
— SPOTTED SAND PIPERS — SNIPE — WOODCOCK — WHITE
IBIS — ROSEATE SPOONBILL — NIGHT HERON — BITTERN —
WHITE EGRET — SNOWY HERON — AMERICAN FLAMINGO.

IN entering upon the consideration of the fifth
order of Birds (Grallatores), the scenes through
which our rambles lay will change materially. The
birds we have thus far described, have, for the most
part, led us to the fields and woods, where we have
marked their graceful motions, flitting from tree to
tree and from grove to grove, or with matchless ease
winging their wild aërial course, high in the vault
of Heaven. But those which we now come to treat
of are mostly the denizens of low marshy grounds,
the borders of streams and lakes, and the shores of
the Ocean,—localities which their peculiar formation
fits them to inhabit. With a few exceptions, a long
bill and a pair of long legs, and a correspondingly
long neck, are the prominent characteristics of this
order.

The first family which we shall notice is that of
the Rail. These birds frequent most of the low
grounds bordering on streams and lakes, both inland

American Flamingo.

and near the coast. There are numerous varieties of the Rails, the most common of which are the Sora Rail and the Virginia Rail. They are both more or less abundant, during the summer months, as far northward as Massachusetts, but retire to the Southern States and Mexico to winter. The flight of these birds during their migrations is swift and long continued, and is performed with a constant beating of the wings. At other times they seem to possess but little activity, except in the use of their legs; their flight being slow and heavy, with the legs dangling, and seldom prolonged to any great distance. The Sora Rail, if pursued by the sportsman, after being forced to rise several times, will at last dive under the water and secrete itself beneath floating weeds, with its bill only above the surface. Respecting this bird Audubon says: "The most curious habit or instinct of this species is the nicety of sense by which they can ascertain the last moment they can remain at any of the feeding grounds at which they tarry in Autumn. One day, you may see or hear the Soras in their favorite marshes, you may be aware of their presence in the dusk of evening; but when you return to the place early next morning, they are all gone. Yesterday the weather was mild, to-day it is cold and raw; and no doubt the Soras were aware that a change was at hand, and secured themselves from its influence by a prompt movement under night."

The plumage of the Rails, although plainly colored, is very soft and compact, particularly on the breast.

18 *   O.

This is very observable in the Virginia Rail, the feathers forming a thick, close, and almost impervious covering. protecting it from the water, in which it not only wades to a considerable depth, but also swims with great ease. This bird is extremely active upon its feet, and upon a level run would almost be

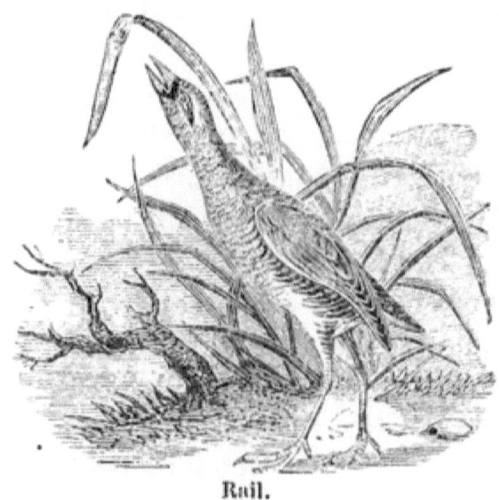

Rail.

a match for a man. If pursued by a dog it will run for a short distance and then tack about, or will rise upon the wing, and with dangling legs fly some dozen yards or so, and then dropping among the grass, scamper off as fast as possible. At the approach of danger it will sometimes cling to the stems of the weeds below the surface of the water, among which it seems almost as much at home as when nimbly skipping about over the broad leaves of the Water Lily which abounds in our inland ponds.

The nest of this Rail is placed on a small elevation formed by collecting together the stalks of a large bunch of grasses; in the centre of this is arranged a quantity of dry weeds to the depth of several inches; upon this slight bed the eggs are deposited, generally four or five in number. The young, when first hatched, are covered with a soft black down, and soon learn to follow the hen through the wet meadows, and upon the sound of danger to enter the water fearlessly.

The food of these birds consists of aquatic insects, snails, worms, crustacea, and the seeds of various grasses which abound in the marshes where they reside. Their habits are partially nocturnal, as they feed both by night and day.

The families and species composing the order Grallatores are so numerous, that it would be impossible, in the limits assigned to this work, to give even a slight glance at the habits of any considerable portion of them; we must, therefore, passing over many familiar and interesting species, confine ourselves to some of the most prominent, and such as will most clearly illustrate the peculiar manners of the Waders.

Late in the Autumn, when the chilling blasts from the regions of eternal snow are beginning to be felt in more southern latitudes, bringing with them myriads of the summer visitors to an Arctic climate, vast trains of ducks, geese, etc., to seek again their winter resorts beneath a milder sky.—then may be heard in the vicinity of our inland lakes and streams the harsh voice of the Whooping Cranes, as they pass swiftly overhead, in companies of from ten to fifty. While migrating they fly high in the air, but when near the spot where they purpose to search for food, they gradually descend, wheeling around in circles over the place until they reach the ground. Here they present a graceful and elegant appearance, the old birds in particular being stately and beautiful objects. The plumage is mostly of a snowy whiteness, except the primaries and the primary coverts, which are nearly black. This bird is quite unknown as a resident or even a transient visitor in the Eastern and Middle States, its haunts being confined to the South and West. It winters as far south as Mexico, and breeds from Oregon northward to the Arctic regions.

Their food consists of the roots of plants, which they dig up with great labor from the mud of shallow ponds which have dried up during Summer; they also resort to the plantations of sweet potatoes, and dig among the hills for the few roots which may have been left in the ground by the farmer. They will also feed on small reptiles, such as frogs, toads, lizards, and even small snakes.

They are said to be extremely wary birds, and very difficult to approach, the least rustling of leaves or the cracking of a stick under foot being sufficient to alarm them, although they may be at a considerable distance. Their sense of sight and hearing is so keen, that they will hear the approach of a hunter at a great distance, and will discover him long before he can see them. When once aware of his advances, no matter how cautious he may be, they will generally prove too much for him, eluding all his attempts to gain access to them.

Prominent among the many attractive objects which may engage the attention of the young naturalist, while tarrying by the sea-side, are those active and beautiful little creatures, the Plovers and Sand Pipers. The species which frequent the whole line of

Sand Piper.

our sea-coast are quite numerous, and the study of their habits would alone afford entertainment and occupation for nearly a whole season. See how beautiful and graceful are their motions as they course along the sand, stopping to examine the shells which the tide in its recess has left upon the beach, or following the retreating breakers to pick up the minute shell-fish borne in by the wave.

Among these we can hardly fail to notice the Ring Plover, Wilson's Plover, and the Piping Plover, — the latter a most beautiful, active, and lovely little bird. It has a sweet, soft, and musical note, which is uttered with a somewhat deceptive effect, and is often heard proceeding from various quarters at the same time, without our being able to discover its source. The flight of this bird is extremely swift, and there are few of its kind that are fleeter of foot. It will run in a straight line before you with such speed that it requires a keen eye to follow it. The nest of the Piping Plover consists merely of a small hole scooped out of the sand, often near the base of a tuft of grass. The female lays four eggs, which are mostly hatched by the warmth of the sand, acquired by exposure to a hot sun. The female, however, always sits upon them by night and during rough weather. The young leave the nest immediately upon being released from the shell, and run about with great activity; and upon the approach of danger they squat so close to the ground, which they very much resemble in color, that it is difficult to discover them.

Although the Plovers are generally abundant on all our Atlantic coasts, yet their haunts are by no means confined to such localities. Many which frequent the sea during the Spring, retire far inland to breed, and some species are seldom known upon the coast. Of those which inhabit our meadows and low grounds, we will select the Kildeer Plover, as the most familiar and the most beautiful.

Almost every farm-house can boast of its pair of Kildeers, which may be seen skimming most gracefully over the fields and meadows, repeating their well-known cry of "Kildeer! Kildeer! dee, dee, dee!" At such times their flight is powerful and easy, somewhat resembling that of the smaller Hawks. Now and then one may be seen following in the track

Kildeer Plover.

of the ploughman, picking up the grubs and worms from the fresh soil. And again you may find him coursing along the shores of some running stream, or upon the muddy banks of a mill-pond, feeding upon the snails or mud-worms which abound in such places. Sometimes it wades into the water to wash and plume its coat, and laying itself down, flutters its wings and splashes about in great glee, until it becomes pretty well soaked, when it retires to a sunny spot to dry.

The nest of this bird is a simple affair, being as a general thing merely a hollow scooped out of the earth, and, when in a wet situation, a few stems are placed around it as a protection. The eggs are four in number, and of a cream color, with markings of brown and black. During the period of incubation, and immediately after the hatching of the young, the old birds manifest much anxiety at the approach of danger. The female endeavors by the usual strata- gem of feigned lameness to entice the intruder away, while the male wheels about overhead in an excited manner, uttering his most earnest entreaties or his most angry reproofs, in hopes no doubt of averting the ruin of his family.

The Kildeer is in every respect a beautiful bird. Whether seen at a distance, sailing or diving with such graceful ease through the buoyant air, or whether upon a nearer view we look upon the lively tints of his exquisite plumage, we cannot but feel that he too is worthy of our notice, and to become the wel- come companion of our rambles.

Among the many active little Sand Pipers to be seen upon our coasts in the Spring and Autumn, are the Red-breasted Sand Piper, the Purple Sand Piper, the Red-backed Sand Piper, and the Semi-palmated Sand Piper. Let us see what we can find out in re- lation to some of them. Of the Red-backed Sand Piper Audubon says: " In Autumn and Winter, this species is abundant along the whole range of our coast, wherever the shores are sandy or muddy, from Maine to the mouths of the Mississippi; but I never found one far inland. Sometimes they collect in flocks of several hundred individuals, and are seen wheeling over the water near the shores or over the beaches, in beautiful order, and now and then so close together as to afford an excellent shot, especially when they suddenly alight in a mass near the sportsman, or when, swiftly veering, they expose their lower parts at the same moment. On such occasions a dozen or more may be killed at once, provided the proper moment is chosen.

"There seems to be a kind of impatience in this bird that prevents it from remaining any length of time in the same place, and you may see it, scarcely alighted on a sand-bar, fly off without any apparent reason to another, where it settles, runs for a few moments, and again starts off on wing. When searching for food they run with great agility, following the retiring waves, and retreating as they advance; probing the wet sands, and picking up objects from their surface, ever jerking up the tail, and now and then uttering a faint cry, pleasant to the ear, and differing from the kind of scream which they emit while on wing."

This bird appears to be an inhabitant of both continents, and although so abundant along the coasts at some seasons, they appear always to retire to the Arctic regions to breed.

The Purple Sand Piper frequents the Atlantic shores from Maine to New York during the Spring and Autumn, but passes the Summer in the Hudson's Bay country. While in the south it seems to prefer rocky shores to the sandy beaches. Their food consists of small shell-fish, worms, and the marine insects which abound among the drifting sea-weeds.

The Semi-palmated Sand Piper is one of those spe-

19

cies whose migrations are not confined to the coast.
Leaving Mexico in the early Spring, these birds
spread themselves eastward along the Gulf and Atlantic shores, and northward by the Mississippi and
other western rivers, making some tarriance in such
situations as are suited to their taste or convenience,
but gradually advancing toward the coasts of Labrador, which appear to be their favorite summer haunts ;
some, however, remaining upon the sea-coast of the
Middle and Southern States during the whole season.

Spotted Sand Piper.

The beautiful and familiar little bird, commonly
known as the Spotted Sand Piper, does not strictly
belong in the same family with the above-named
species, but being very closely allied, we will notice
it here.

During the spring and summer months, all our
rivers, small streams, and ponds, seem to abound with
this active and sprightly creature.   While upon the

ground it appears to be constantly in motion, now darting along the water's edge after a spider, and now dabbling in the mud with its bill in search of worms, all the while wagging its stumpy little tail in a most ludicrous manner; no matter in what position it is seen, except when flying, this perpetual motion of the tail is observable; and even the young acquire the singular habit almost immediately upon leaving the shell. These little fellows also run about with wonderful speed, which no doubt enables them to escape danger with great facility. The old birds manifest great anxiety in protecting them, fluttering about with much concern at the approach of an intruder, using every stratagem they are capable of to secure their escape. The following beautiful incident is related by Wilson:

"My venerable friend, Mr. William Bartram, informs me that he saw one of these birds defend her young for a considerable time from the repeated attacks of a ground-squirrel. The scene of action was on the river shore. The parent had thrown herself, with her two young behind her, between them and the land; and at every attempt of the squirrel to seize them by a circuitous sweep, raised both her wings in an almost perpendicular position, assuming the most formidable appearance she was capable of, and rushed forward on the squirrel, who, intimidated by her boldness and manner, instantly retreated; but, presently returning, was met as before, in front and on flank, by the daring and affectionate bird, who, with her wings and whole plumage bristling up, seemed swelled to twice her usual size. The young crowded together behind her, apparently sensible of their perilous situation, moving backward and forward as she advanced or retreated. This interesting scene lasted for at least ten minutes; the strength of the poor parent began evidently to flag, and the attacks of the squirrel became more daring and frequent, when my good friend stepped forward from his retreat, drove the assailant back to his hole, and rescued the innocent from destruction."

Two of the commonest and best-known birds among the Grallatores are probably the Snipe and the Woodcock. Renowned among the gunners as affording the rarest and most exciting sport, and no

Snipe.

less renowned among the gastronomes of our cities, who love better to indulge their appetites over a well-cooked brace of either, than to apply their energies to the doubtful and difficult task of obtaining them.

The Snipe is familiar only as a transient visitor during Spring and Autumn, its summer haunts being among the cold countries of the north, where it raises its brood and returns to pass the Winter in the south.

The Woodcock is a summer resident in the Northern, Eastern, and Middle States, where it is a very abundant species, frequenting the low grounds and swampy woods of almost every neighborhood. This fact would perhaps be disputed by some in consequence of their not being aware that the habits of the bird are nocturnal, and would not therefore meet the eye of most, unless accidentally disturbed. The early twilight is the signal for the Woodcocks to retire to their cover, and the approach of dusk to sally forth in quest of food; this consists of earth-worms, which they obtain by probing the soft mire with their bills, through which they appear to suck them up without withdrawing their bills from the mud, in the manner of the Curlews and some other water birds. They will sometimes resort to the woodland and scratch among the dry leaves for the worms which are often secreted there; but this probably is only during hot weather, when the marshy places are partly dry, and the supply of food less abundant.

Neither in respect to form nor general appearance can the Woodcock lay claim to beauty or grace. The markings of its plumage are indeed very delicate, but the contrasts of color are less pleasing than in many of its associates. The head, which is rather a shapeless affair, has the appearance of being a con-

19 *

stant burden, and the eyes, which are large, are
placed so high up as to give it quite a singular look ;
but these peculiarities, no doubt, assist it in its nocturnal rambles, the large eye admitting more light,
and its elevated position commanding a greater range
of vision.	Thus, it can discover with greater ease
the approach of an enemy, and while flying over its
favorite feeding grounds, can more readily select a
spot suited to its tastes.

The nest of this bird is loosely built of dry leaves
and grass, and generally placed at the foot of some
low bush, or by the side of a prostrate log, in the
darkest and most secluded part of the woods.	The
eggs are mostly four, and are of a clayish-colored
ground, with irregular patches of brown and purple
thickly sprinkled over the surface.	The young com-
mence to run about as soon as hatched, and so rapid
is their growth, that at the age of six weeks they are
almost as active on the wing as their parents.

The next family of the Waders which we shall notice, is that of the Ibis ; of this group we number
four species, one of which, the richly-colored Scarlet
Ibis, is a very doubtful resident among us, as a few
only have ever been seen in the country, and it seems
likely that its occurrence among us has been purely
accidental, as it is evidently a native of a warm
Southern climate.	It appears to be quite plentiful
in the West India Islands, and in the Bahamas,
which are no doubt its natural haunts.

The White Ibis inhabits the southern parts of
Florida, where it is resident.	In Summer, some in-

dividuals have been seen as far north as New Jersey, but it may be considered rare north of the Carolinas. On some of the islands at the southern extremity of Florida these birds congregate in great numbers to breed. Their nests are placed on the low shrubbery or trees, and are sometimes very close together, Audubon having counted forty-seven on a single plum-tree.

Respecting some of its habits we quote the following from the above-named author: "The flight of the White Ibis is rapid and protracted. Like all other species of the genus, these birds pass through the air with alternate flappings and sailings; and I have thought that the use of either mode depended upon the leader of the flock; for, with the most perfect regularity, each individual follows the motions of that preceding it, so that a constant appearance of regular undulations is produced through the whole line. If one is shot at this time, the whole line is immediately broken up, and for a few minutes all is disorder; but as they continue their course, they soon resume their former arrangement. The wounded bird never attempts to bite or to defend itself in any manner, although, if only winged, it runs off with such speed as often to escape the pursuer.

"At other times the White Ibis, like the Red and the Wood Ibises, rises to a great height in the air, where it performs beautiful evolutions. After they have thus, as it were, amused themselves for some time, they glide down with astonishing speed, and alight either on trees or on the ground. Should the sun be shining, they appear in their full beauty, and the glossy black tips of their wings form a fine contrast with the yellowish white of the rest of their plumage.

"The manner in which this bird searches for its food is very curious. The Woodcock and the Snipe, it is true, are probers as well as it, but their task requires less ingenuity than is exercised by the White or Red Ibis. It is also true that the White Ibis frequently seizes on small crabs, slugs, and snails, and even at times on flying insects; but its usual mode of procuring food is a strong proof that cunning enters as a principal ingredient in its instinct. The cray-fish often burrows to the depth of three or four feet in dry weather, for before it can be comfortable it must reach the water. This is generally the case during the prolonged heats of Summer, at which time the White Ibis is most pushed for food. The bird, to procure the cray-fish, walks with remarkable care toward the mounds of mud which the latter throws up in forming its hole, and breaks up the upper part of the fabric, dropping the fragments into the deep cavity that has been made by the animal. Then the Ibis retires a single step, and patiently waits the result. The cray-fish, incommoded by the load of earth, instantly sets to work anew, and at last reaches the entrance of its burrow; but the moment it comes in sight, the Ibis seizes it with his bill."

In the localities where the Ibis abounds may also be seen the graceful form and beautiful colors of that singular bird, the Roseate Spoonbill. It is much to be regretted that so many of the most beautiful water birds should be confined in their rambles to the southern extremity of our country. How nicely would this noble and elegant bird decorate the Fauna of our Northern and Middle States! It is not likely, however, that any are to be found much to the northward

Roseate Spoonbill.　　　　　Night Heron.

of the lower parts of Georgia; its principal haunts being near the shores of the Gulf of Mexico, and the extensive bayous and inlets which abound in the vicinity. Let us imagine ourselves upon one of those beautiful islands or keys which skirt the southern

P

coast of the Evergreen State. Amidst a dense growth
of Cactus, with its sharp and rigid spines everywhere
menacing our steps; a wide-spread expanse of water
is before us, whose surface is as lovely and tranquil
as the sky that overshadows it; here and there the
tall stems of the graceful palm-trees are reflected
upon its bosom. In this secluded spot the sight of a
flock of these birds may frequently be enjoyed, and,
if well concealed from their view, we may study
their manners at our leisure. Standing with their
wings partly extended, in the bright rays of the sun
they present a beautiful spectacle, the deep roseate
tints upon the sides and upon the wings being then
displayed to the finest advantage. Behold them moving about, with measured tread and stately attitude,
upon the muddy shore, or wading into the shallows
to search for food. Here their broad spoon-like bills
are brought into energetic action. Thrusting the
head and sometimes the neck into the water or mire,
and seizing upon the various small shell-fish, insects,
and other water animals, they literally chew them up
with their powerful bills before swallowing them.
After feeding awhile, they will all indulge in a wild
sally into the free air, ascending sometimes to a considerable height, moving about in the most graceful
manner, crossing and recrossing each other, and performing a great variety of interesting aërial evolutions; then the whole flock suddenly return to their
feeding grounds, plunging through the air with great
power and speed.

Associated with the Spoonbills will be found a great

variety of another class of Waders, called Herons,
which are not only much more abundant, but more
widely distributed, — many of the species extending
their migrations as far to the north as the State of
Maine. Among those with which our readers are
most likely to be familiar, are the Night Heron, or
Qua-Bird, the Bittern, the Great White Egret, and
the Snowy Heron, or Little Egret. A full-plumaged
male Night Heron is unquestionably a beautiful bird.
Standing about two feet in height, its head crowned
with a loose, flowing crest of elongated feathers of a
shining green of the deepest shade, from the centre
of which project three slender feathers, pure white,
and about eight inches in length, each having its
edges so rolled up as to make it a perfect tube. The
upper part of the back and the scapulars are of a
deep blackish green, the wings grey, with a shade of
lilac. The throat is pure white, which gradually
shades into a light cream color upon the breast and
whole lower parts.

Except during the breeding season, this is a shy
and wary bird, and extremely difficult to approach.
While a flock is engaged in feeding, one of their
number acts as sentinel, to give the alarm at the
least sound of danger. This is a common practice
with many birds of this class, and it is said that the
Spoonbills feed with great confidence when in company with Herons, taking warning at the voice of
their sentinel. The Night Heron may be examined
at leisure, and even shot in great numbers, by secre-
ting oneself near the spot where they regularly roost

by day. Here, as they arrive singly or a few at a
time, a good opportunity is afforded the naturalist
to study some of their habits. In the selection of a
breeding place, they generally assemble in small companies of from twenty to fifty, and appropriate a
clump of cedars, cypress, or mangrove, according
to the locality which they inhabit, where their nests
sometimes crowd the branches to within a few feet
of the ground. These Heronries are mostly upon
the borders of some stagnant pools or in the vicinity
of cedar, cypress, and other swamps, as well as upon
the shores of those sea-islands which are covered
with evergreens. The nests are large, and irregu-
larly formed of sticks placed one above another, to
the height of a few inches; their structure is sometimes so slight as to tumble to pieces before the young
are fit to fly. These birds, when once in possession
of a breeding place suited to their tastes, will return
to it annually, and repair the old nests, until circumstances force them to abandon it.*

The Great White Egret is another of those elegant
and stately birds with which our water scenery is
often beautified. Along the banks of our great rivers, and sometimes of our smaller streams and mill-
ponds, groups of these fairy-looking creatures may
frequently be seen, wading at their leisure among
the tall reeds and other plants which abound in the
shallow water. Here, with untiring patience, they
move about slowly and cautiously, awaiting the appearance of some unlucky fish, or water animals of

* Audubon.

almost any kind. If it is possible to approach them sufficiently near to observe their motions while thus occupied, we shall hardly fail to be gratified with the sight. Here is one fine fellow, standing over three feet and a half in height. He has straightened up his tall and graceful figure to its full extent, and is looking around suspiciously, but not observing any danger, he composes himself to his work. What a noble bird! His plumage, of snowy whiteness, fairly glistens in the sun's rays; and the long, flowing plumes, which form a train of exquisite delicacy, are waving in the gentle breeze. Now with silent watchfulness he intently eyes the quiet water, his neck curved so as to bring the head to rest above the shoulders. In this position he stands motionless as a statue, engaged either in quietly contemplating what is going on around him, or perhaps in watching for fresh game. Let us now apprise them that we are too near for their convenience. Suddenly the whole troop spread their broad wings, and in the most majestic manner move slowly away. For a long distance we can watch them; their heads drawn in to the shoulders, the long legs extended to their utmost in the rear, like a rudder, and their ample wings beating the air in slow and measured strokes. This showy bird appears to inhabit the whole line of the Atlantic States as far as Massachusetts, confining itself principally to the vicinity of those waters which flow toward the sea, seldom, if ever, being found very far in the interior.

The Little Egret, or Snowy Heron, is another of

20

those birds which are always conspicuous for the perfect whiteness of their plumage; but of all the species, this is probably gifted with a coat of the most

Snowy Heron.

delicate and beautiful texture. The head is ornamented with a long, flowing crest, composed of fine thread-like plumes. Upon the lower part of the neck the feathers are lengthened, and hang down in what might be called a loose beard-like tuft, while from the upper part of the back proceed a number of long slender plumes of lace-like delicacy, extending over the rump and turning upward at the extremity, the fine filaments hanging from the shafts like the hair from the tail of a bobtailed horse.

This beautiful bird seems to give the preference to the salt marshes, which line the coast from Maine to Florida. Here, during the breeding season, they are generally abundant; and, as is the custom of the Herons, their nests are clustered together in communities of greater or less numbers, according to circumstances. In New Jersey, the cedars which gen-

erally skirt the low grounds near the shore, are selected as their resort. The nests are placed sometimes two or three upon the same tree, but seldom
more. In whatever position they build, it is said
that the nests always front the water, and very often
overhang it. These communities seem very social in
their disposition, living upon good terms with the
Night Herons, Green Herons, and Grakles which
have their nests near by.

We cannot close our notices of the Grallatores without a brief description of that gorgeously plumaged
bird, the American Flamingo. Although extremely
rare, and seldom seen within our territory except
upon the most southern extremity of Florida, and
upon the little islands which skirt its coast, it seems
entitled to a place among those which annually visit us
from the south.

This elegant bird is about four feet in height, and
is wholly of a bright scarlet color, with the exception
of the primaries and a part of the secondaries, which
are black. Its habits are very similar to those of the
Waders in general; its flight consists of alternate sailing and flapping of the wings, the neck and legs being both extended to the utmost. The nest of the
Flamingo is a curious structure; it is built in the
midst of the shallow water of some salt-pond, the
mud being heaped up into a pile about two or three
feet high, on the top of which a hollow is scooped
out, where the female lays two white eggs about the
size of those of a goose. In covering the eggs during incubation, she is obliged to stand with one foot

in the water, her body being supported by the nest. The Flamingo, like its neighbors the Herons, is exceedingly shy and difficult to approach ; when moving over the water, it generally flies low, but upon nearing land, unless its purpose is to alight, it immediately ascends to a considerable height, as though to escape danger. We clip the following from Audubon's notes respecting this bird :

" On the 7th of May, 1832, while sailing from Indian Key, one of the numerous islets that skirt the southeastern coast of the Peninsula of Florida, I for the first time saw a flock of Flamingoes. It was on the afternoon of one of those sultry days which, in that portion of the country, exhibit toward evening the most glorious effulgence that can be conceived. The sun, now far advanced toward the horizon, still shone with full splendor, the ocean around glittered in its quiet beauty, and the light fleecy clouds that here and there spotted the heavens, seemed flakes of snow margined with gold. Our bark was propelled almost as if by magic, for scarcely was a ripple raised by her bows as we moved in silence. Far away to seaward we spied a flock of Flamingoes advancing in 'Indian line' with well-spread wings, outstretched necks, and long legs directed backward. Ah ! reader, could you but know the emotions that then agitated my breast ! I thought I had now reached the height of all my expectations, for my voyage to the Floridas was undertaken in a great measure for the purpose of studying these lovely birds in their own beautiful islands. I followed them with my eyes, watching as it were every beat of their wings; and as they were rapidly advancing toward us, Captain D. A. Y., who was aware of my anxiety to procure some, had every man stowed away out of sight, and our gunners in readiness. The pilot, Mr. Egan, proposed to offer the first taste of his 'groceries' to the leader of the band. He was a first-rate shot, and had already killed many Flamingoes. The birds were now, as I thought, within a hundred and fifty yards; when suddenly, to our extreme disappointment, their chief veered away, and was of course followed by the rest. Mr. Egan, however, assured us that they would fly round the Key, and alight not far from us, in less than ten minutes; which in fact they did, although to me these minutes seemed almost hours. 'Now they come,' said the pilot; 'keep low.' This we did; but, alas! the Flamingoes were all, as I suppose, very old and experienced birds, with the exception of one; for on turning round the lower end of the Key, they spied our boat, again sailed away without flapping their wings, and alighted about four hundred yards from us, and upward of one hundred from the shore, on a 'soap-flat' of vast extent, where neither boat nor man could approach them."

20 *

CHAPTER XII.

NATATORES.

THE Natatores comprises a large variety of Geese,
Swans, Ducks, Gulls, Tern, and all other web-footed
birds, except the Flamingo, which, notwithstanding
it has this peculiarity, we have placed among the
Grallatores, its habits and manners, and general appearance, agreeing more nearly with them than with
the Natatores. It, however, appears to be a connecting link between the two, the form of the bill and
the mode of feeding being similar to that of the
Duck tribe; while its long legs, stately attitude, its
wading propensities, and other prominent characteristics, must ever associate it with the Heron and
other kindred families.

The limits of this work will not admit of our entering into any very extensive description of the
many beautiful and interesting objects which will
present themselves to our view as we examine the
field before us. Without intending to slight our
web-footed friends, we shall therefore select for description a few honest representatives from the various

families which compose the order, giving slight notices to others as occasion may offer.

We will commence with the Canada Goose, a very abundant species in the Northern, Middle, and Western States, at some seasons of the year. During

Canada Goose.

Summer, these birds seek the more remote district of Labrador, where they breed,—returning, however, at the first approach of cold, and distributing themselves throughout a vast extent of country to the southward during Winter.

The habits of this bird are quite interesting, and

from a most graphic description of them given by
Audubon, we glean the following particulars:

"The general spring migrations of the Canada
Goose may be stated to commence with the first melt-
ing of the snows in our Middle and Western districts, or from the 20th of March to the end of April;
but the precise time of its departure is always determined by the advance of the season; and the vast
flocks that winter in the great savannas or swampy
prairies southwest of the Mississippi, such as exist
in the Opelousas, on the borders of the Arkansas
river, or in the dismal 'Everglades' of the Floridas,
are often seen to take their flight, and steer their
course northward, a month earlier than the first of
the above-mentioned periods.

"It is my opinion that all the birds of this species,
which leave our States and Territories each Spring
for the distant north, pair before they depart. This,
no doubt, necessarily results from the nature of their
place of summer residence, where the genial season
is so short as scarcely to afford them sufficient time
for bringing up their young and renewing their plumage, before the rigors of advancing Winter force
them to commence their flight toward milder countries. This opinion is founded on the following
facts:— I have frequently observed large flocks of
Geese, in ponds or marshy grounds, or even on dry
sand-bars, the mated birds renewing their courtships
as early as the month of January, while the other
individuals would be contending or coquetting for
hours every day, until they all seemed satisfied with

the choice they had made; after which, although
they remained together, any person could easily perceive that they were careful to keep in pairs.

"Such are the conflicts of these ardent lovers, and
so full of courage and of affection toward their females are they, that the approach of a male invariably ruffles their tempers as well as their feathers. No
sooner has the goose laid her first egg, than her bold
mate stands almost erect by her side, watching even
the rustling sound of the breeze. The least noise
brings from him a sound of anger. Should he spy
a raccoon making its way among the grass, he walks
up to him undauntedly, hurls a vigorous blow at him,
and drives him instantly away. Nay, I doubt if man
himself, unarmed, would come off unscathed in such
an encounter.

"The Canada Goose is less shy when met with far
inland, than when on the sea-coast. They usually
feed in the manner of swans and fresh water ducks,
that is, by plunging their heads toward the bottom
of shallow ponds or the borders of lakes and rivers,
immersing their fore parts, and frequently exhibiting
their legs and feet with the posterior portion of their
body elevated in the air. They never dive on such
occasions. Wherever you find them, and however
remote from the haunts of man the place may be,
they are at all times so vigilant and suspicious, that
it is extremely rare to surprise them. In keenness
of sight and acuteness of hearing, they are perhaps
surpassed by no bird whatever. They act as senti-
nels toward each other, and during the hours at which

the flock reposes, one or more ganders stand on the watch. At the sight of cattle, horses, or animals of the deer kind, they are seldom alarmed, but a bear or a cougar is instantly announced; and if on such occasions the flock is on the ground near water, the birds immediately betake themselves in silence to the latter, swim to the middle of the pond or river, and there remain until danger is over. So acute is their sense of hearing, that they are able to distinguish the different sounds or footsteps of their foes with astonishing accuracy. Thus the breaking of a dry stick by a deer is at once distinguished from the same accident occasioned by a man. If a dozen of large turtles drop into the water, making a great noise in their fall, or if the same effect is produced by an alligator, the Wild Goose pays no regard to it; but however faint and distant may be the sound of an Indian's paddle, that may by accident have struck the side of his canoe, it is at once marked, every individual raises its head and looks intently toward the place from which the noise has proceeded, and in silence all watch the movements of their enemy."

Of the Swan family we have two species, the American Swan and the Trumpeter Swan. The latter appears to be exclusively a western species, being most abundant in the vicinity of the Mississippi, Missouri, and other western rivers, during Winter, and breeding from California northward to the fur countries. The American Swan is found in Winter along the Atlantic coasts, sometimes in considerable numbers, particularly in Chesapeake Bay, but appears to be scarce south of this, its principal haunts being to the northward. During the summer months the shores of the Polar Sea afford it a safe retreat, where it may rear its young in comparative safety

American Swan.

The flight of these birds is powerful and rapid, and is often prolonged to a wonderful extent. During their migrations they soar to a great height, overtopping the mountains, and seldom pause during the journey between our latitude and the place of their summer abode, except when their progress is impeded by a storm, above the region of which they mostly travel. They always advance in small flocks in the shape of a V, the leader being at the point. When they arrive at the place of their destination, which is generally at night, they occupy themselves at once in making amends for their long abstinence from food, and join in a wild chorus of congratulations which almost makes the shores ring. While feeding, or during the operation of dressing and arranging their plumage, they are apt to be very noisy, their notes varying much from high to low, according to circumstances. But so vigilant are they, that upon the least note of alarm from the sentinel all is immediately quiet, and they move noiselessly away from the scene of danger.

Mallard Duck.

Of the Duck tribe we have a large number of species, many of them possessed of beautiful plumage and interesting habits. Quite prominent among these is the Common Mallard, with its stately head of rich golden green, and back and breast and wings of varied shades of brown and blue and black and white. From this fine bird has sprung many of the races of Domestic Ducks which are now dispersed over the country. But in his wild state he bears so little resemblance to his degenerate progeny, that one would scarcely recognize his connection with it. The Mallard is found in most parts of the country during the winter season, except in the Eastern States. Audubon says they "generally arrive in Kentucky and other parts of the western country [from the north], from the middle of September to the first of October, or as soon as the acorns and beech-nuts are fully ripe. In a few days they are to be found in all the ponds that are covered with seed-bearing grasses. Some flocks, which appear to be guided by an experienced leader, come directly down on the water with a rustling sound of their wings, that can be compared only to the noise produced by an Eagle in the act of stooping upon its prey; while other flocks, as if they felt uneasy respecting the safety of the place, sweep around and above it several times in perfect silence, before they alight. In either case, the birds immediately bathe themselves, beat their bodies with their wings, dive by short plunges, and cut so many capers that you might imagine them to be stark mad. They wash themselves and arrange their dress, before commencing their meal; and in this, other travellers would do well to imitate them.

"Now, toward the grassy margins they advance in straggling parties. See how they leap from the water to bend the loaded tops of the tall reeds. Woe be to the slug or snail that comes in their way. Some are probing the mud beneath, and waging war against the leech, frog, or lizard that is within reach of their bills; while many of the older birds run into the woods, to fill their crops with beech-nuts and acorns, not disdaining to swallow also, should they come in

21 Q

their way, some of the wood-mice that, frightened by the approach of the foragers, hie toward their burrows. The cackling they keep up would almost deafen you, were you near them, but it is suddenly stopped by the approach of some unusual enemy, and at once all are silent."

During the autumn months our inland streams and lakes mostly abound with many varieties of Ducks, of forms and degrees of beauty as numerous as their species. We would gladly give our readers a full description of these bright wanderers, but our limits will allow only of a few remarks respecting most of them, while with some of the most interesting we may spend more time.

Wood Duck.

We have already become a little acquainted with the Mallard, both as the occupant of our private duck-ponds, and also as a denizen of the free air; let us now see if we cannot find something to interest us in that model of beauty of its kind, the Common Summer or Wood Duck. This is one of the few species which remain within the limits of the States throughout the year, much the larger proportion retiring to the "far north" to breed. The Summer Duck is certainly one of the most elegant of its tribe; its plumage being richly glossed with green and gold, and purple and black, in some places mottled with white, or finely barred with black and fawn. The head presents a fine appearance, surmounted by a long crest of green, and the cheeks beautifully marked with black and white. It appears to be widely spread over the whole extent of the country, from Louisiana to Maine, and westward some distance up the Missouri river. Within these limits it may almost be said to be a constant resident. It generally builds its nest in a hollow tree, frequently in the deserted hole of a large Woodpecker, giving the preference to such trees as are near the water, or which overhang pools or marshes. The number of eggs which the female deposits varies much; Audubon says from six to fifteen; Wilson speaks of a nest containing thirteen. It is a singular fact, according to the first-named author, that upon the female having completed her number of eggs, she is at once deserted by the male, who, joining with a few others, roams about until the young are able to fly, when the old and young unite in one flock, and remain together until another season comes round.

The Green and the Blue-winged Teal are also two handsome Ducks, but are only known to us as transient visitors in the spring and autumn months, the cold regions of the fur countries being their usual place of resort during Summer.

The Canvass-back is the famous Duck which is generally considered by epicures as the finest of all the Duck family,—its flesh being thought to possess a peculiarly agreeable flavor, which no other fowl can claim. The most common winter resort of these celebrated Ducks is the Chesapeake Bay and the rivers and streams belonging to it, such as the Susquehanna, Patapsco, Potomac, and James rivers. Here they sometimes assemble in flocks of such great numbers as to cover the surface of the water for acres in extent, and when they rise suddenly the noise of their wings resembles thunder. The abundance of their favorite food (a species of Valisneria), a grass-like plant which grows to the height of a few feet above the water, the roots of which seem to form their main sustenance, is evidently the great attraction for these birds, as of later years their numbers appear to have decreased, while at the same time the plant has become less abundant. These Ducks are often seen feeding in company with several other species, such as the Black-headed Duck, the Widgeon and the Red-headed Duck. They all appear to live upon the same plant; the Canvass-back and the Black-head diving to obtain the roots, while the Widgeon and the Red-head prefer the leaves. The Canvass-back has also been found on the waters of the Hudson, and upon some of the western rivers; but its chief winter haunts are to the southward, while its summer life is passed far away to the north.

Leaving the Canvass-backs in company with the Ring-necks and Ruddy Ducks, the beautiful Bird Duck, and the Velvet Duck, with its coating of black, we pass on to the well-known Eider Duck. This

Eider Duck.

elegant bird, which inhabits the northern portions of both continents, must, for various reasons, be looked upon with great interest by the student of Nature; and the value of its down, as a promoter of ease and comfort, must claim for it equal celebrity with the Canvass-back. In some localities their nests are usually built upon rocky precipices which over-hang the ocean, and are lined with the soft down which the female plucks from her breast. In those countries where this down is collected as an article of commerce, in order to increase the quantity pro-duced in one season, the nest is deprived of its eggs

21*

as well as the down; the female again plucks her bosom, and lays a fresh complement of eggs, which are also taken; a third time she makes the effort to raise a brood, when the male sometimes assists in lining the nest by taking the down from his own breast. This brood they are allowed to raise, for, if their hopes of progeny are entirely destroyed, they will abandon the place; whereas, if once attached to a spot, they return to it year after year with their young.

The Eider Duck is seldom found south of the vicinity of New York. Further north and to the eastward as far as the bay of Fundy, it becomes more abundant; and to Labrador thousands of pairs, it is said, annually resort to breed and spend the short Summer. Respecting their habits in these countries, Audubon says: " In Labrador, the Eider Ducks begin to form their nests about the last week of May. Some resort to islands scantily furnished with grass, near the tufts of which they construct their nests; others form them beneath the spreading boughs of the stunted firs, and in such places, five, six, or even eight, are sometimes found beneath a single bush. Many are placed on the sheltered shelvings of rocks a few feet above high-water mark, but none at any considerable elevation; at least none of my party, including the sailors, found any in such a position. The nest, which is sunk as much as possible into the ground, is formed of sea-weeds, mosses, and dried twigs, so matted and interlaced as to give the appearance of neatness to the central cavity, which rarely exceeds seven inches in diameter. In the beginning of June the eggs are deposited, the male attending upon the female the whole time. The eggs, which are regularly placed on the moss and weeds of the nest, without any down, are generally from five to seven, three inches in length, two inches and one-eighth in breadth, being thus much larger than those of the Domestic Duck, of a regular oval form, smooth-shelled, and of a uniform pale olive-green. When the full complement of eggs has been laid, she begins to pluck some down from the lower parts of her body; this operation is daily continued for some time, until the roots of the feathers, as far forward as she can reach, are quite bare, and as clean as a wood from which the undergrowth has been cleared away. This down she disposes beneath and around the eggs. When she leaves the nest to go in search of food, she places it over the eggs; and in this manner, it may be presumed to keep up their warmth, although it does not always ensure their safety, for the Black-backed Gull is apt to remove the covering, and suck or otherwise destroy the eggs. The care which the mother takes of her young for two or three weeks, cannot be exceeded. She leads them gently in a close flock in shallow waters, where, by diving, they procure food; and, at times, when the young are fatigued, and at some distance from the shore, she sinks her body in the water, and receives them on her back, where they remain several minutes."

The Long-tailed Duck is another beautiful species,

which breeds away to the northward, and visits us in great numbers during Winter, being found almost everywhere on the Atlantic coast. They are a noisy, lively species, and owing to their reiterated cries, they have been called "Noisy Ducks;" they have, however, other names applied to them, such as "Old Wives" and "Old Squaws."

Hooded Merganser.

With the Hooded Merganser we must close our brief notices of the Ducks. This showy and elegant bird is more an inhabitant of our western and southern waters than of the eastern coast. It cannot then be said to be an abundant species in Pennsylvania. It breeds along the Mississippi, the Ohio, and the great Lakes, as well as further northward, and during Winter it is said sometimes to retire as far southward as Mexico. The plumage of this bird is indeed very beautiful. The thick, flat, tufted crest which covers the whole head, and much resembles a hood, gives it a sprightly and animated appearance. This crest, together with the whole head, neck, breast, and upper part of the back, are singularly marked with black and pure white, which is well contrasted with the rich brown of the sides and flanks. The female is a much plainer bird, but not without some claims to beauty.

Like the Common Wood Duck, the Merganser seems to prefer placing its nest in some hollow tree, to building, as most other species do, upon the ground. The eggs are deposited on a bed of dried weeds, feathers, and some down from the breast of the bird. When the young are hatched, they are conveyed to the water by the parent, who gently takes them in her bill, and removes them one by one to their favorite element. Here she leads them among the tall grass and weeds, and teaches them to procure the snails and insects that come within reach. The Hooded Merganser is an expert diver, and in this way often escapes the sportsman's gun, plunging, almost in a twinkling, below the surface, on the first intimation of danger.

With the name of the Pelican most of our readers are familiar, while with its appearance they may be wholly unacquainted. The American White Pelican, which Audubon is pleased to style a "splendid bird," but which is quite too awkward to merit that term, is rarely seen in the middle districts, while to the north and west and south it seems to be more common. According to Dr. Richardson,* it is abundant

---

* Author of "Fauna Boreali Americana."

in the fur countries, flying about in dense flocks all Summer. To these parts, and to the Rocky Mountain districts, it mostly resorts for the purpose of breeding,—its winter quarters extending southward from Carolina to Texas, along the coasts as well as inland.

The plumage of this bird is quite white, except a portion of each wing, which is nearly black. From the back part of the head hangs a short crest of loose feathers. This crest, together with a tuft of feathers on the breast, is of a pale yellow color, as is also the pouch which hangs from the lower mandible. The upper mandible is armed at a short distance from the extremity with a sharp bony process, which occupies about one-fourth its length. The Pelicans are apt to assemble in flocks of considerable size, and resort to the same feeding ground, where they will arrange themselves on the margin of some sand-bar, pluming themselves, and preparing for the coming meal. During this time, should one of them gape, all, as if by sympathy, open their long and broad mandibles, yawning lazily and ludicrously. At length hunger compels their return to the water. With awkward gait they waddle along as though they were out of their element; but when they reach the water's edge they seem like other creatures. How beautifully do they float upon the surface as they arrange themselves for their work! The following paragraph from Audubon shows their manner of taking food: "In yonder nook, the small fry are dancing in the quiet water, perhaps in their own manner bidding farewell to the orb of day, perhaps seeking something for their supper. Thousands there are, all gay, and the very manner of their mirth, causing the waters to sparkle, invites their foes to advance toward the shoal. And now the Pelicans, aware of the faculties of their scaly prey, at once spread out their broad wings, press closely forward with powerful strokes of their feet, drive the little fishes toward the shallow shore, and then with their enormous pouches spread like so many bag-nets, scoop them out, and devour them in thousands."

We must now spend a little time among the large and interesting families of the Terns and Gulls, and watch their beautiful motions as they skim over the surface of the ocean, now rising upon the bosom of the gale, and now with the swiftness of an arrow plunging into the deep in pursuit of their prey.

The Black Skimmer, or Shearwater, is a very singular bird, inhabiting our southern sea-coasts, where, during most of the night, in localities which it frequents, its hoarse cry may be heard as it sails over the water in search of food. With wide-spread wings it swiftly glides along, the lower mandible ploughing the water, while the upper mandible, which is movable, is elevated a little above it. In this manner it secures its prey, sometimes rising above the surface, and again dipping its great bill as fresh objects appear. Thus, the whole night long, with almost untiring energy, it skims the surface of the deep, winging its graceful and buoyant flight beneath the light of the pale moonbeams, until day dawns, when it betakes itself to the beach or some sand-bar to rest.

There are perhaps few of our readers who have the opportunity of visiting any part of our extensive sea-coast during Summer, who can fail to notice two birds; these are the Common Tern and the Least Tern. They are so abundant, and their beautiful motions so attractive, that the most unobservant must pause to watch and admire them. They differ from each other principally in size, the former being much the larger. Their

Arctic Tern.

plumage is quite similar, being mostly of a snowy-white, tinged on the back with light blue-grey, while a patch of black covers the crown of the head. Swallow-like in their form, they seem to mimic in their motions the antic gambols of that gay and nimble little bird,—skimming with sylph-like ease over the white-capped breakers, watching intently for their prey, upon which they dart almost with the swiftness of thought. The Least Tern is particularly social, and seeming to possess a degree of confidence in man, which perhaps he little deserves, he approaches him fearlessly, flying about him with the most unsuspicious familiarity. We would recommend every visitor at the sea-coast to study the habits of these two lovely birds.

Along the shores of Maine, Nova Scotia, or of Labrador, the Arctic Tern is seen gambolling in the air above the voyager, whose eye is riveted upon its graceful evolutions. Now it sweeps over some solitary green isle, — then, amidst the floating icebergs, stoops to pick up some hapless shrimp. Little care is required to construct its nest, which is generally on a low sand-bank or desert island; and in a short time the little Terns burst the shell, hobble toward the water, and soon are on the wing, far out at sea. The first snow-storm from the Polar lands, however, drives before it multitudes of these sprightly and daring rovers, to a southern clime.

This bird is occasionally seen upon the Jersey shore in Autumn, whence it departs in early Spring. Some follow the windings of the coast up to Newfoundland, while others, younger and perhaps more fearful, fly inland, passing along the St. Lawrence to the Magdalene islands and the "ice-bound" Labrador.

Audubon remarks that when a female Arctic Tern has been killed and floats upon the water, her mate will alight upon and caress her, as if she were still living. He tried the experiment several times, and invariably with the same result.

A curious fact may be stated here, in reference to this genus, — that all the Terns that breed in the

22

northern parts of the United States, and in the Polar
regions, sit closely on their eggs; while the species
that breed in more southern latitudes incubate only
during the night or in rainy weather.

Of the family of Gulls, so well known and so
widely diffused, we notice first the species bearing
the name of Bonaparte, in allusion to the well-known
naturalist. This bird is found at times in great
numbers along our sea-board, from the Bay of Fundy,
and even higher latitudes, to the coast of Florida.
It has also been observed sweeping over the Ohio
river, in search of small fishes or floating garbage.
When examined after death, the stomachs are found
to contain shrimp, young fishes, fatty substances, and
sometimes coleopterous insects. In Spring, when
the shad enter the bays and rivers to deposit their
spawn, this Gull begins to show itself, as if for the
purpose of preying upon the shoals, which, however,
is not the case. It is described as being very gentle
in some localities, scarcely heeding the presence of
man.

The Great Black-backed Gull, the largest of the
tribe, delights in sailing over the rugged crags of
Labrador. He moves in wide circles, with loud,
harsh cries, far above the multitudes of smaller birds
below, who instinctively dread the approach of this
tyrant, or prepare to defend their young broods from
its powerful beak. The fish sink deeper as he approaches, while the other Gulls fly as fast as possible
from their enemy. At length he spies, perhaps, the
carcass of a whale, and, with fierce cries, darts down

upon the putrid mass.   Tearing, tugging, and swallowing piece after piece, until he is surfeited, he lies down exhausted; but, owing to the great digestive power of his stomach, in a short time he is again on the wing to some well-known isle, where thousands of young birds or eggs are to be found.   There, without remorse at the screams of the parents, he begins leisurely to break open and devour, until he has again satisfied his craving appetite.   But though so tyrannical, he is yet a coward, and sneaks off at the approach of the Skua, a much smaller but bold sea-bird, which is always ready to attack the relentless robber.

Upon the western shores of Labrador, for an extent of three hundred miles, this king of Gulls is found in great numbers in the breeding season.   Toward the commencement of Summer they arrive one by one, the older ones first, greeting with loud cries the first sight of their native land.   With many bows and gesticulations the pairing proceeds, until, at the right time, they fly off to one of the many desert isles that line the shore, and build their nests beneath a projecting shelf, or in a wide cleft of a rock.   They are formed of moss and sea-weeds, carefully arranged, being two feet in diameter, five or six inches in height, and lined with feathers and dry grass.   Not more than three eggs are ever laid in one nest, which, like those of most other Gulls, afford good eating.   When the young are five or six weeks old, they take to the water, uttering the same sounds as the old birds. Even at that early period they show great greediness in eating. If a dead duck or even one of their own species is thrown to them, they tear it in pieces, drink the blood, and swallow the flesh in large morsels, each one trying to rob the other of his share. They will attempt to take down codlings ten inches in length, and, though the shape of the fish may be distinctly traced along the neck, and the birds are evidently suffering from the pressure on the windpipe, they will not disgorge their prey. They will attack flocks of young Ducks while swimming beside their mother, when the latter takes wing, and the frightened brood dive. If among the bushes, they are safe; but if no shelter is near, they are likely to be caught by their voracious enemy. The Eider Duck is the only one that offers resistance to save her young; but when sitting on eggs in any open situation, the Gull will drive her off and suck them before her eyes. He will sometimes seize flounders on the edge of the shallows, but not being able to swallow them whole, flies to some rock, and beats the fish until it can be torn to pieces. The stomach of this bird appears to be capable of reducing feathers, bones, and other hard substances, with ease.

The whole length is nearly thirty inches, and a full-grown specimen will weigh three pounds. The fishermen and settlers of Newfoundland and Labrador kill large numbers of the young ones when nearly able to fly, and, after skinning them, salt them down for food.

We turn now to a bird familiarly known to sailors, the world over. Constantly flapping its wings, and showing only a single spot of white, its dusky form darts in every direction along the swelling waves. Never fatigued, the tiny Petrels seldom alight, though they seem now and then to walk upon the foaming crests of the water. When the gale approaches, and the little wanderers are unable to bear up against its

Wilson's Petrel. (Mother Carey's Chicken.)

fury, they retreat for shelter to the stern of the nearest vessel, where they remain until the blue sky overhead again tempts them to fly forth to pick up the floating fragments on the sea. There are three varieties of this interesting bird. We will notice but one, Wilson's Petrel, which is sometimes confounded with a smaller kind, both being familiarly called Mother Carey's Chickens. Of their migrations but little is known. The range is stated by Audubon not to extend eastward beyond the Azores, nor lower than

22 *   R

the Gulf of Mexico. This species breeds on low sandy banks, scantily covered with grass, and called Mud Islands, off the southern extremity of Nova Scotia. In the middle of Summer they form burrows to the depth of two feet, place at the bottom a few bits of dry grass, and lay only one egg. In two months the young follow their parents to sea, and are scarcely distinguishable from them.

With its wings nearly at right angles with the body, in calm weather the Petrel runs or rather hops upon the water, patting it with its feet, and keeping its head downward in search of small fishes. Now and then the ear is attracted by its note, resembling the syllables "Kee-re-kee-kee!" which are more frequently uttered at night than by day. In every clime the sailor regards with friendly interest this lively and sociable creature. When, flitting over the long ocean swell, they chase one another in play, every one hails them as harbingers of fair weather; but when dull moaning sounds are heard afar, and the little rovers sweep near the vessel, or cluster near its sheltering sides, they give the timely warning to close haul the sails before the tall masts creak and tremble in the gale.

Having now reached the limit assigned to our volume, we must pass by without notice several other species of water birds found upon our coasts, whose habits are interesting; but we trust that the perusal of what has been written will have so far instructed and pleased our readers as to stimulate them to a more extended study of the delightful subject of Ornithology.

We trust also that the acquaintance thus formed with the sweet Songsters of the Wood and Field, will be a means of tuning hearts to praise the Great Creator of every living thing, Our Father who is in Heaven.

THE END.